Nonlinear Dynamics in Cardiovascular Signals

Nonlinear Dynamics in Cardiovascular Signals

Guest Editor

Claudia Lerma

MDPI

Basel • Beijing • Wuhan • Barcelona • Belgrade • Novi Sad • Cluj • Manchester

Guest Editor
Claudia Lerma
Departamento de Biología
Molecular
Instituto Nacional de
Cardiología Ignacio Chávez
Mexico City
Mexico

Editorial Office
MDPI AG
Grosspeteranlage 5
4052 Basel, Switzerland

This is a reprint of the Special Issue, published open access by the journal *Entropy* (ISSN 1099-4300), freely accessible at: https://www.mdpi.com/journal/entropy/special_issues/189OYPH31W.

For citation purposes, cite each article independently as indicated on the article page online and as indicated below:

Lastname, A.A.; Lastname, B.B. Article Title. *Journal Name* **Year**, *Volume Number*, Page Range.

ISBN 978-3-7258-4449-4 (Hbk)
ISBN 978-3-7258-4450-0 (PDF)
https://doi.org/10.3390/books978-3-7258-4450-0

Contents

entropy

MDPI

Article

Nonlinear Dynamics of Heart Rate Variability after Acutely Induced Myocardial Ischemia by Percutaneous Transluminal Coronary Angioplasty

Martín Calderón-Juárez [1,2], Itayetzin Beurini Cruz-Vega [1], Gertrudis Hortensia González-Gómez [3] and Claudia Lerma [2,*]

[1] Plan de Estudios Combinados en Medicina, Faculty of Medicine, Universidad Nacional Autónoma de México, Mexico City 04510, Mexico; martin.cal.j@comunidad.unam.mx (M.C.-J.); ibcruzve@comunidad.unam.mx (I.B.C.-V.)

[2] Department of Electromechanical Instrumentation, Instituto Nacional de Cardiología Ignacio Chávez, Mexico City 04480, Mexico

[3] Department of Physics, Faculty of Sciences, Universidad Nacional Autónoma de México, Mexico City 04510, Mexico; hortecgg@ciencias.unam.mx

* Correspondence: dr.claudialerma@gmail.com

Abstract: Several heart rate variability (HRV) characteristics of patients with myocardial ischemia are associated with a higher mortality risk. However, the immediate effect of acute ischemia on the HRV nonlinear dynamical behavior is unknown. The objective of this work is to explore the presence of nonlinearity through surrogate data testing and describe the dynamical behavior of HRV in acutely induced ischemia by percutaneous transluminal coronary angioplasty (PTCA) with linear and recurrence quantification analysis (RQA). Short-term electrocardiographic recordings from 68 patients before and after being treated with elective PTCA were selected from a publicly available database. The presence of nonlinear behavior was confirmed by determinism and laminarity in a relevant proportion of HRV time series, in up to 29.4% during baseline conditions and 30.9% after PTCA without statistical difference between these scenarios. After PTCA, the mean value and standard deviation of HRV time series decreased, while determinism and laminarity values increased. Here, the diminishment in overall variability caused by PTCA is not accompanied by a change in nonlinearity detection. Therefore, the presence of nonlinear behavior in HRV time series is not necessarily in agreement with the change of traditional and RQA measures.

Keywords: heart rate variability; recurrence plot; surrogate data; nonlinearity; percutaneous transluminal coronary angiography; reperfusion

1. Introduction

Heart rate variability (HRV) has proven to be a sensitive tool for risk stratification in patients after acute myocardial infarction (AMI) [1]. In general, decreased time-domain indices are related to higher cardiovascular and all-cause mortality [1,2]. More recently, nonlinear measures have been considered to identify high-risk patients [3]. The nonlinear approach, in addition to clinical applications [4], has been implemented to study cardiovascular and autonomic physiology in experimental settings [5].

Nonlinear behavior is often assumed in physiological time series, such as HRV. However, it is appropriate to investigate whether the nonlinear properties at issue are justified by the data [6], as this has implications in the use of methods from the nonlinear approach, particularly in the interpretation of the results. Often, a diminishment in nonlinear measures through aging and pathologic conditions is taken as an indicator of "loss" in nonlinear behavior, although this does not mean that nonlinear dynamic cease to exist in the time series [7].This issue was addressed with Fourier transform-based surrogate data in patients about ten days after AMI [8], in which nonlinear dynamics were identified in HRV using entropy-based measures of time series complexity and regularity.

One of the major drawbacks of Fourier transform-surrogate data testing is that nonlinearity will be detected because the time series are either nonlinear or nonstationary [9]. To overcome this issue, it has been proposed to use methods based on wavelet decomposition to generate surrogate data that preserve nonstationary behavior of time series [10,11]. We have implemented such methods in short-term HRV of healthy subjects and in pathologic conditions [12], in which we propose to study nonlinear behavior through recurrence quantification analysis (RQA). This approach is suitable for short, noisy, and nonstationary time series [13], properties that are present in HRV.

As mentioned above, nonlinearity has been assessed several days after AMI [8], and during the myocardial ischemia [14,15]. Furthermore, revascularization therapy carried out by percutaneous transluminal coronary angioplasty (PTCA) may induce relevant changes in HRV dynamics. HRV linear modifications have been suggested as indicators for improved autonomic regulation during PTCA [16] and during follow-up after PTCA [17–19]. However, the HRV dynamics and the presence of nonlinearity have not been assessed by the RQA approach in patients with acutely induced occlusion of major coronary arteries.

The aim of the study is to assess the presence of nonlinearity in HRV time series from patients during acutely induced myocardial ischemia by PTCA, as well as evaluate HRV through traditional and RQA measures.

2. Materials and Methods

2.1. Study Design and Data Collection

Data collection was obtained from the "STAFF III database" on Physionet [20,21], which includes data from 104 patients who were elective patients for percutaneous coronary intervention (PCI) where single prolonged balloon inflation was introduced in one of the major coronary arteries according to the case. This database was obtained between 1995–1996 at Charleston Area Medical Center (Charleston, WV, USA), containing standard 12-lead ECG recordings.

ECGs in two different moments (lead V) were obtained during the PCI before and after the intervention in two established places: the patient's room and the catheterization laboratory. For this work pre-inflation (baseline) recordings were obtained immediately before any catheter insertion, and the post-inflation ECG recordings were obtained immediately after the balloon deflation. In both moments, the ECGs were acquired for 5 min during supine position in the catheterization laboratory.

According to database information, there were 20 patients with prior myocardial infarction (MI). To minimize the artifacts and noise from skeletal muscle, the Mason-Likar electrode configuration was used. Data acquisition was performed with a custom-made equipment by Siemens-Elema AB (Solna, Sweden). The ECG was digitized at a sample rate of 1000 Hz with 0.625 μV of amplitude resolution. The study protocol was approved by the Research and Ethics Committee of our institution (protocol number 22–1309).

From the initial sample of 104 patients, during signal processing, QRS detection, and RR beat correction, 36 patients were excluded due to having total recording time < 300 s (n = 1), ECG artifacts, arrhythmias, or sustained ectopic beats (n = 11), not having either the pre-inflation or post-inflation recording in the catheterization laboratory room (n = 12), reading errors in ECG data (n = 7), or having >5% of RR intervals replaced due to arrhythmias with adaptive filtering procedure [22] (n = 5). Finally, 68 subjects were considered for the analysis (40% were female patients). Table 1 shows the clinical characteristics of the studied sample. When comparing the subgroup by sex, both age and proportion of patients having prior MI were similar between groups. All patients with prior MI from the initial sample were included.

Table 1. Clinical characteristics from 68 patients with myocardial ischemia who underwent a percutaneous transluminal coronary angioplasty (PTCA) procedure. Results are shown as mean ± standard deviation or absolute value (percentage).

Variable	All (N = 68)	Female (N = 27)	Male (N = 41)	*p* Value
Age (years)	59 ± 12	59 ± 10	59 ± 13	0.996
Prior MI	20 (29%)	6 (22%)	14 (34%)	0.291

MI: myocardial infarction.

2.2. ECG Processing and QRS Complex Identification

The ECG recordings of 5 min length were visually inspected by three trained observers. Additionally, correct identification of R waves was visually supervised in the collaborative platform PhysioZoo V1.5.6 software under peak detector "rqrs" allowing manual correction of miss-detections, artifacts, and arrhythmias [23]. An adaptive filter was applied to correct ectopic heartbeats replacing them by normal RR interval values [22]. Heartbeat replacement was applied in 39 recordings pre- (71.4%) and post-inflation (28.6%), considering beat correction of 0.55% as the limit to include. Hence, we refer to NN intervals as the time between "normal" heartbeats [24].

2.3. Heart Rate Variability Analysis

2.3.1. Time-Domain and Frequency-Domain Measures

The mean value of 5-min HRV time series was calculated (meanNN [ms]), its standard deviation (SDNN [ms]), and the standard deviation of the difference between consecutive NN intervals (SDSD [ms]) [24]. Before estimation of the frequency-domain measures, NN time series were resampled at 3 Hz, then a 300-data point. Hamming window was applied through the resampled data with 50% overlap. Finally, discrete Fourier transform was applied, and we calculated the low (*LF* (ms^2), 0.04–0.15 Hz) and high frequency (*HF* (ms^2), 0.15–0.4 Hz) bands. These measures are also expressed as normalized units (n.u.), as follows [24]:

$$LF(\text{n.u.}) = \frac{LF}{Total\ power - VLF}, \quad (1)$$

$$HF(\text{n.u.}) = \frac{HF}{Total\ power - VLF}, \quad (2)$$

where *VLF* is the very low frequency band (<0.04 Hz). We also show the *LF* (ms^2)/*HF* (ms^2) ratio.

2.3.2. Recurrence Quantification Analysis

The embedding parameters for each time series: embedding dimension (m) and embedding delay (τ) were determined by the first local minimum at zero in the false nearest neighbors function and the first local minimum at averaged mutual information function, respectively.

After embedding, the recurrence plot construction is defined in [13]:

$$R_{i,j} = \Theta\left(\varepsilon_i - \|\vec{x_i} - \vec{x_j}\|\right),\ i,j = 1,\ldots,N, \quad (3)$$

where a tolerance to define a recurrence ε_i of the point x_j in the vicinity centered in x_i corresponds, in this case, to the fixed amount of neighbors (FAN) norm (recurrence density was set to = 0.07). If the distance between two points falls within the vicinity, the Heaviside function $\Theta(x)$ assigns $R_{i,j} = 1$, and said point is considered a recurrence point, otherwise $R_{i,j} = 0$. Then a matrix of these binarized values is created and is graphically represented as recurrence plot, where black points correspond to $R_{i,j} = 1$, and white points to $R_{i,j} = 0$.

The MATLAB toolbox "Cross Recurrence Plot" developed by Marwan et al. [25] (https://tocsy.pik-potsdam.de/CRPtoolbox/, accessed on 29 January 2023) was used to estimate the nonlinear measures determinism (DET) and laminarity (LAM), both measures useful to characterize the autonomic control of HRV in humans, i.e., larger DET and LAM values reflect vagal withdrawal [26]. DET is the proportion of recurrence points that form diagonal structures (Equation (4)) and LAM is the proportion of recurrence points that form vertical structures (Equation (5)), calculated with minimal diagonal $l_{min} = 2$ and vertical length $v_{min} = 2$, based on the histogram $P(l)$ of diagonal and vertical structures, respectively.

$$DET = \frac{\sum_{l=l_{min}}^{N} lP(l)}{\sum_{l=1}^{N} lP(l)} \tag{4}$$

$$LAM = \frac{\sum_{v=v_{min}}^{N} lP(l)}{\sum_{v=1}^{N} lP(l)} \tag{5}$$

2.4. Surrogate Data Testing

We applied the algorithm pinned wavelet iterative amplitude adjusted Fourier transform (PWIAAFT) [10] developed by Keylock, C. (https://sites.google.com/site/chriskeylocknet/software, accessed on 29 January 2023), which preserves nonstationary behavior, opposed to Fourier transform-based surrogates that introduce stationarity on surrogate data [9] and may be prone to mistake nonstationarity for nonlinearity. Briefly, in this method, maximal overlap discrete wavelet transform (MODWT) is used to decompose the original time series, then a threshold (ρ) is established to fix a set of wavelet coefficients. On the remaining wavelet scales, the iterative amplitude adjusted Fourier transform (IAAFT) [9,11,27] is applied. Additionally, the algorithm fits a cubic Hermitian polynomial in the fixed coefficients before applying IAAFT in these [11], also referred to as gradual wavelet reconstruction. A flow chart describing the IAFFT algorithm is shown in Figure 1 and the wavelet-based algorithm is described with the flow chart of Figure 2, both procedures are described in detail in [11].

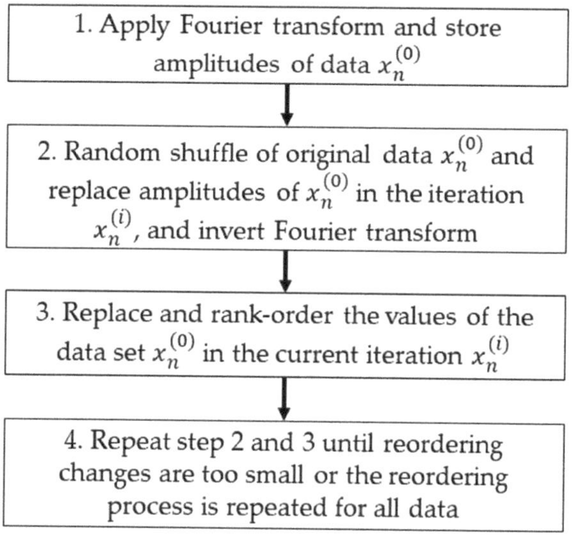

Figure 1. Flow chart of the iterative amplitude adjusted Fourier transform (IAAFT) algorithm [9,11].

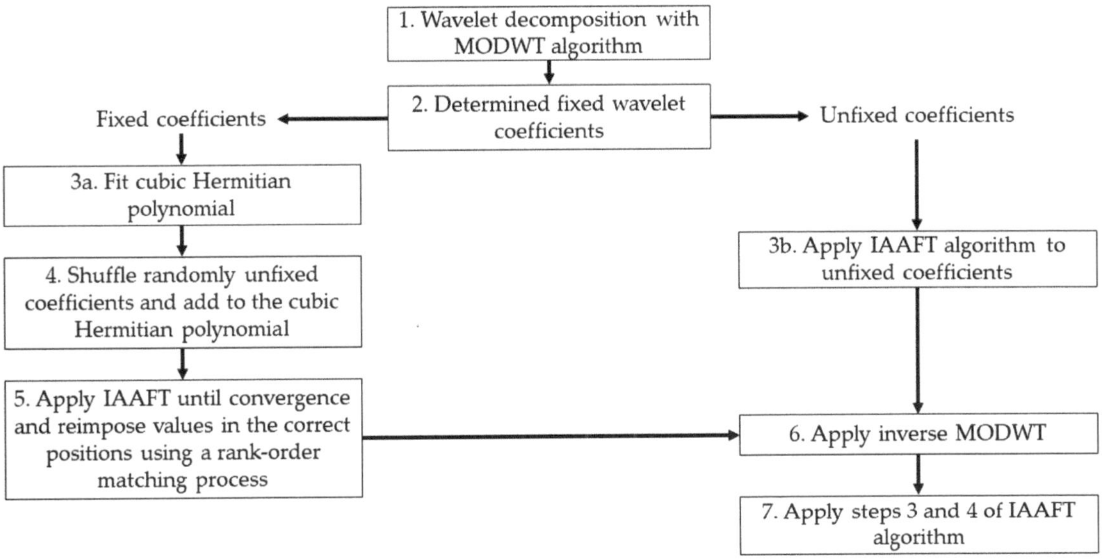

Figure 2. Flow chart of the pinned wavelet iterative amplitude adjusted Fourier transform (PWIAAFT) algorithm (gradual wavelet reconstruction) algorithm. A full description of the algorithm is in [11].

This method was used in 5-min HRV time series in a previous work [12] where no benefit was found from using a threshold for wavelet fixation $\rho > 0.01$. Therefore, in this work we use $\rho = 0.01$.

A group of 99 surrogates is generated, and a discriminative nonlinear statistic (in this case, LAM and DET) is measured in the set of surrogates and the original time series. If the value measured on the original data lies beyond the fifth percentile of either side of the statistical distribution curve, this time series is classified as nonlinear.

2.5. Statistical Analysis

Categorical clinical variables are expressed as absolute frequency (relative frequency, in brackets) and were compared with the Pearson's chi-square test. Continuous data had normal distribution (Kolmogorov–Smirnov test, $p > 0.05$) and are shown as mean ($\pm$standard deviation). Continues data were compared with paired Student's t-test or analysis of variance (ANOVA) for repeated measures with post-hoc comparisons adjusted by the Bonferroni method, considering prior MI as the comparison factor between groups and measurement before and after PTCA as the comparison factor within groups. The indices LF(ms^2), HF(ms^2), and LF/HF were log transformed. We calculated 95% confidence interval for the percentage of nonlinear time series classified as nonlinear (Clopper–Pearson method). The statistical analysis was performed by Statistical Package for the Social Sciences (SPSS) version 21.0 (IBM Corp., Armonk, NY, USA).

3. Results

Table 2 shows the results of HRV linear indices in all participants (N = 68). There are statistically significant differences in meanNN, SDNN, SDSD, ln(LF(ms^2)), and ln(HF(ms^2)) between the two moments, being smaller after the PTCA procedure compared with the baseline. LF (n.u.), HF (n.u.), and ln (LF/HF) had no significant difference when compared before and after PTCA.

Table 2. Heart rate variability indexes from 68 patients with myocardial ischemia who underwent a percutaneous transluminal coronary angiography (PTCA) procedure. Results are shown as mean ± standard deviation.

	Before PTCA	After PTCA	*p* Value
	Time-domain measures		
meanNN (ms)	882 ± 149	856 ± 134	0.044
SDNN (ms)	33.9 ± 17.9	23.8 ± 13.1	<0.001
SDSD (ms)	18.8 ± 13.9	11.6 ± 9.3	<0.001
	Frequency-domain measures		
ln(LF (ms^2)) [1]	5.107 ± 1.358	4.044 ± 1.438	<0.001
ln(HF (ms^2)) [1]	3.818 ± 1.397	2.697 ± 1.388	<0.001
LF (n.u.)	74.2 ± 17.8	75.4 ± 17.2	0.580
HF (n.u.)	25.9 ± 17.8	25.1 ± 17.0	0.683
ln(LF/HF) [1]	1.289 ± 1.084	1.348 ± 1.035	0.660

[1] Natural logarithm (ln) was used for log transformation. meanNN: mean of NN intervals. SDNN: standard deviation of NN intervals. SDSD: standard deviation of the difference between consecutive NN intervals. LF: low frequency. HF: high frequency.

Table 3 shows the analysis of linear HRV indices before and after PTCA in patients grouped by prior MI (N = 20) or no prior MI (N = 48). While meanNN, SDNN, and SDSD decreased significantly after PTCA only in the group that had no history of prior MI, ln(LF(ms^2)) and ln(HF(ms^2)) decreased significantly after PTCA in both groups. LF (n.u.), HF (n.u.), and ln(LF/HF) had no significant change after PTCA in both groups. In all linear HRV indices, there were no significant differences between groups (prior MI vs. no prior MI) for both before and after PTCA.

Table 3. Heart rate variability indexes from patients who underwent PTCA grouped by history of prior myocardial infarction (MI). Results are shown as mean ± standard deviation.

	Prior MI (N = 20)		No Prior MI (N = 48)	
	Before PTCA	**After PTCA**	**Before PTCA**	**After PTCA**
meanNN (ms)	864 ± 147	850 ± 130	891 ± 152	860 ± 137 *
SDNN (ms)	27.6 ± 12.8	23.2 ± 10.8	36.5 ± 19.2	24.1 ± 14.0 **
SDSD (ms)	14.2 ± 7.9	9.6 ± 5.6	20.7 ± 15.4	12.4 ± 10.5 **
ln(LF (ms^2))	4.846 ± 1.358	4.195 ± 1.450 *	5.216 ± 1.357	3.981 ± 1.444 **
ln(HF (ms^2))	3.462 ± 1.287	2.646 ± 1.153 **	3.966 ± 1.427	2.718 ± 1.485 **
LF (n.u.)	75.783 ± 15.800	77.741 ± 17.568	73.490 ± 18.715	74.399 ± 17.093
HF (n.u.)	24.268 ± 15.832	22.259 ± 17.568	26.510 ± 18.715	26.082 ± 16.874
ln(LF/HF) [1]	1.385 ± 1.060	1.549 ± 1.136	1.249 ± 1.103	1.264 ± 0.990

[1] Natural logarithm (ln) was used for log transformation. meanNN: mean of NN intervals. SDNN: standard deviation of NN intervals. SDSD: standard deviation of the difference between consecutive NN intervals. LF: low frequency. HF: high frequency. * $p < 0.05$ (before vs. after PTCA, same group). ** $p < 0.01$ (before vs. after PTCA, same group).

Figure 3 shows an example of the HRV time series and recurrence plot before PTCA (left column). The panels on the right column correspond to an illustrative surrogate time series and recurrence plot that, overall, exhibits a similar texture to the original. Figure 4 shows an example of the original and surrogate HRV time series after PTCA in the same fashion as Figure 3.

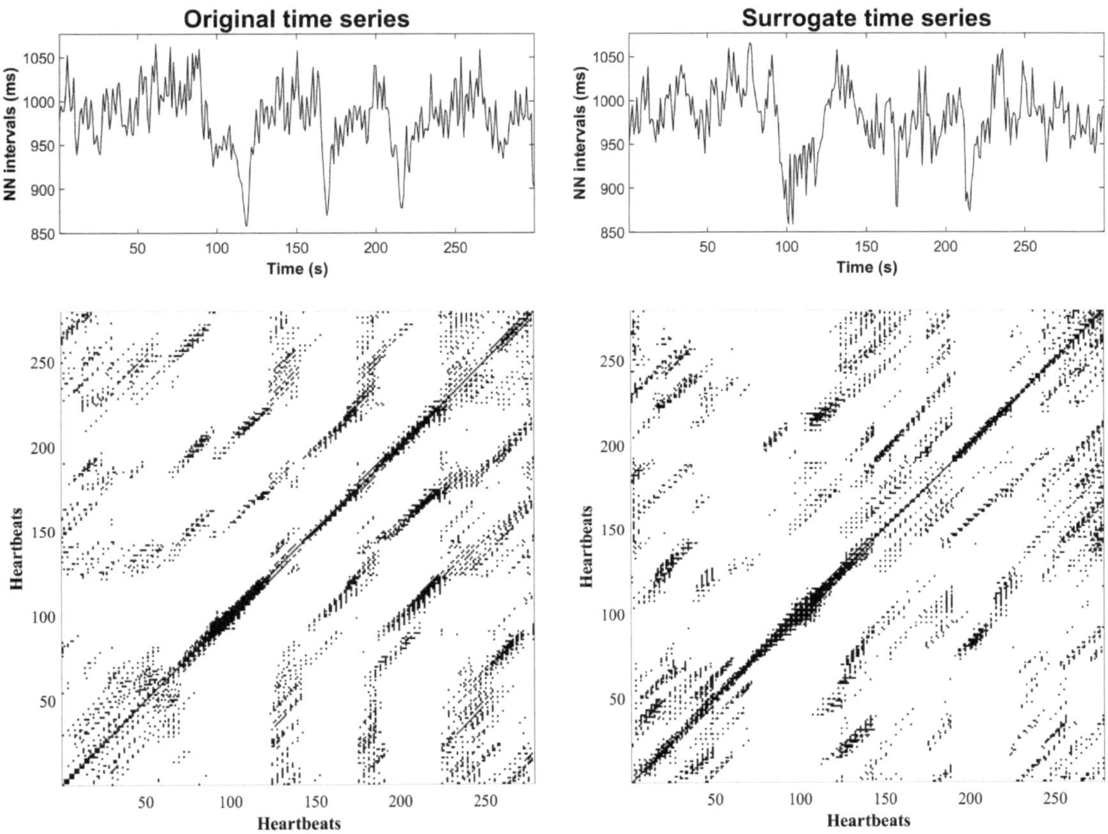

Figure 3. Illustrative examples of an original HRV time series and the corresponding recurrence plot before PTCA (**left** column), and a surrogate HRV time series with the corresponding recurrence plot (**right** column).

Figure 5 shows the percentage of HRV time series classified as nonlinear before and after PTCA with determinism (DET) and laminarity (LAM) as discriminative nonlinear statistics. The percentage of nonlinear HRV time series is as follows: (a) DET, before PTCA 29.4%, after PTCA 30.9%; (b) LAM, before PTCA 26.5%, after PTCA 23.5%. No statistically significant differences were found before and after PTCA with either of the nonlinear measures.

DET and LAM values were significantly higher after PTCA compared with baseline condition (Figure 6); DET (before [0.666 ± 0.164] vs. after [0.739 ± 0.158] PTCA), LAM (before [0.687 ± 0.192] vs. after [0.761 ± 0.165] PTCA). When comparing DET and LAM of patients grouped by history of prior MI, the increase in determinism and laminarity was significant in both groups, and there were no significant differences between groups (both before and after PTCA) (Figure 7). It can be observed in both Figures 6 and 7 that the values in each patient can increase or decrease after PTCA, rather than moving their values in the same direction. In Figure 7, the plots of SDNN and SDSD were added to illustrate that also in linear HRV indices, individual direction of change after PTCA may vary between subjects in both groups, and yet only in the groups that had no prior history of MI the increase in SDNN and SDSD was significant.

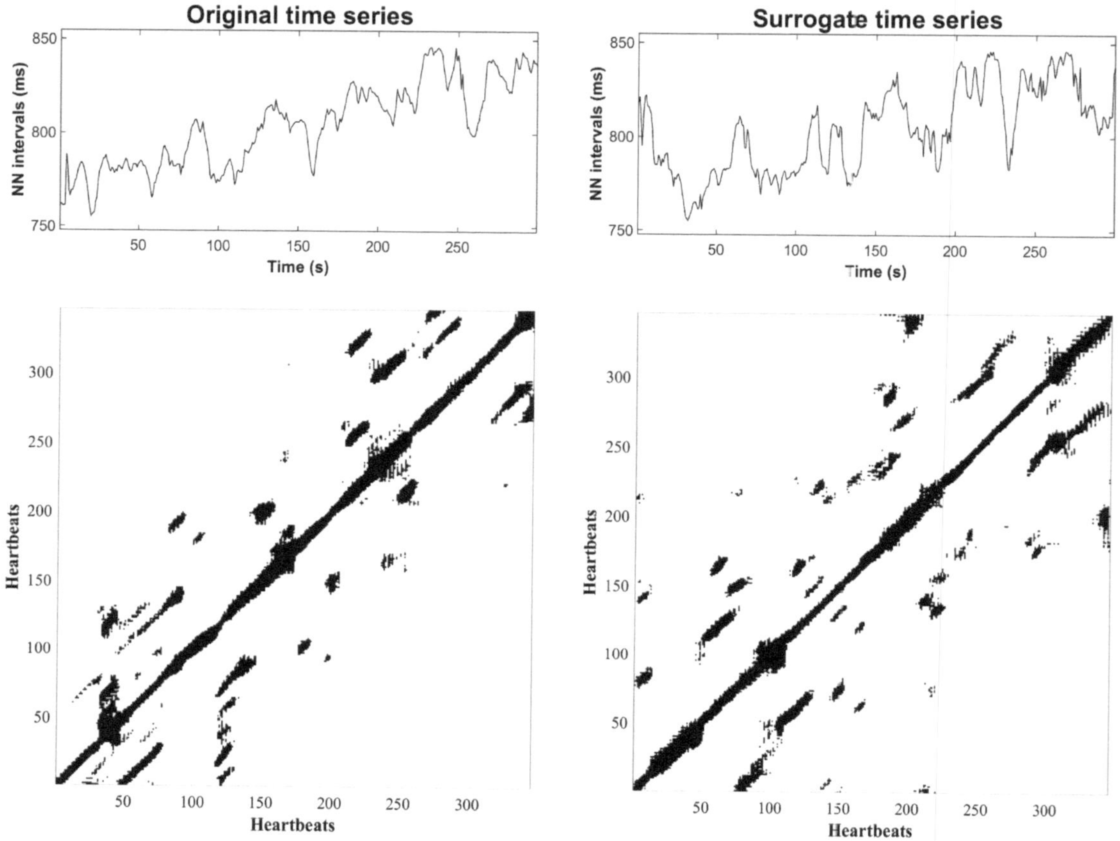

Figure 4. Illustrative example of and original HRV time series and recurrence plot after PTCA (**left** column). Surrogate HRV time series and recurrence plot (**right** column).

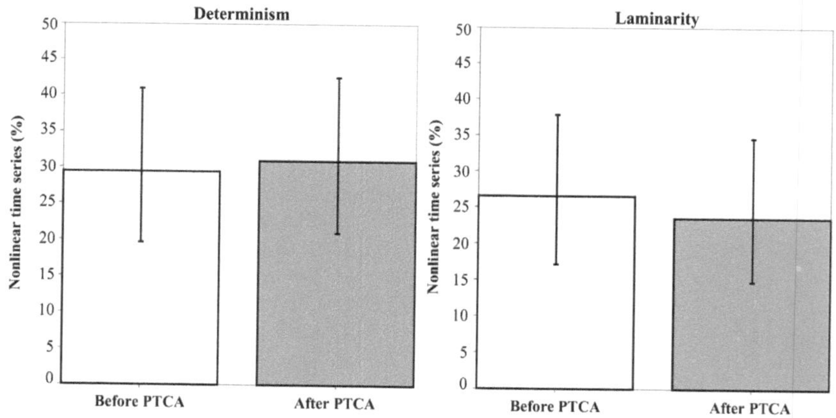

Figure 5. Percentage (95% confidence interval) of nonlinear HRV time series in patients with myocardial infarction (N = 68) before and after PTCA. Determinism (DET) and laminarity (LAM) were used as discriminative nonlinear statistics.

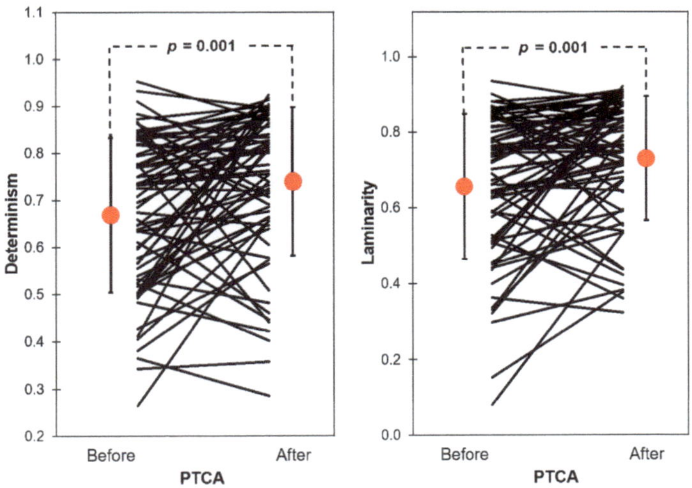

Figure 6. Determinism and laminarity in patients with myocardial infarction (N = 68) before vs. after PTCA, mean (red dot) ± 1 SD is shown.

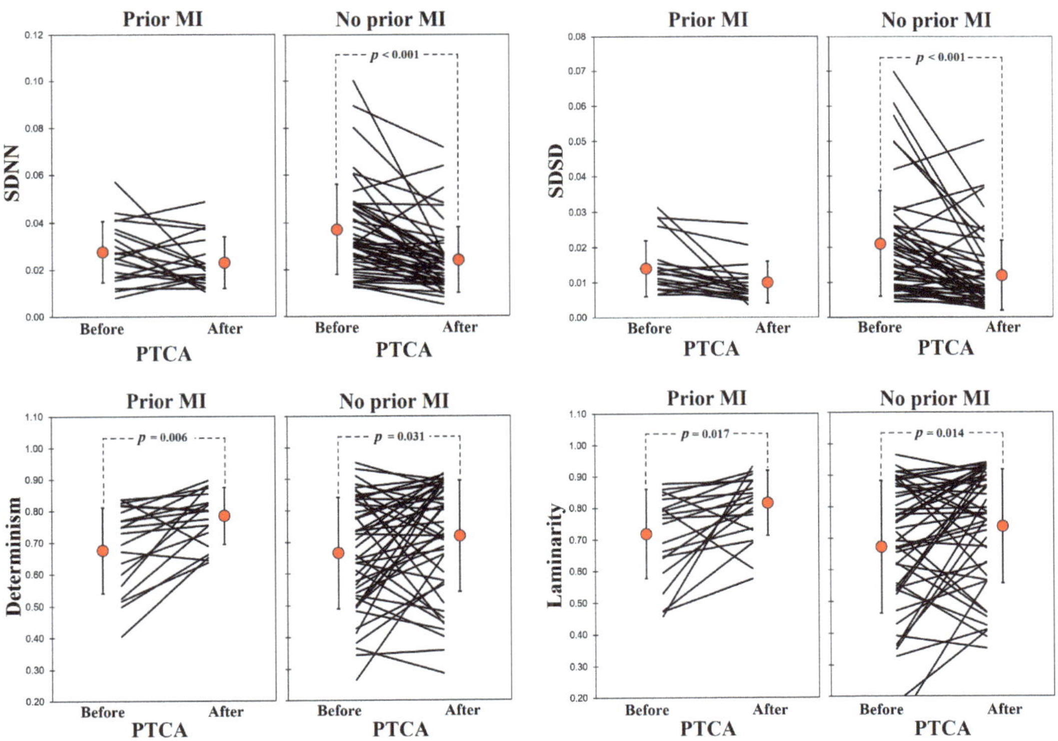

Figure 7. SDNN, SDSD, DET, and LAM in patients before and after PTCA grouped by history of prior myocardial infarction (MI). Mean (red dot) ± 1 SD is shown.

4. Discussion

In this work, we tested the presence of nonlinear information in HRV time series of patients with acutely induced ischemia during elective PTCA. We show that the overall statistical variation of HRV time series diminishes immediately after PTCA; also, the DET and LAM values are higher after the revascularization procedure.

In patients undergoing PTCA, SDNN diminished during the first hour after the reperfusion procedure and eventually increased beyond their baseline conditions (before PTCA). Moreover, the mean of RR intervals gradually increased during the next 24 h after PTCA [28]. Other authors, who measured SDNN before and 24 h after PTCA found that this value increases; however, they did not observe any relevant change in SDSD [29]. In the present study, we found a decrease in meanNN, SDNN, and SDSD right after the PTCA. These results reflect that the setpoint of HRV (meanNN) decreases in response to revascularization (i.e., mean heart rate increases), and the gross variability of the time series is narrower (SDNN and SDSD). Moreover, lower SDNN in patients treated with PTCA before hospitalization discharge is related with an increased risk of major clinical events (death or readmission for a new AMI) [30]. Also, we found that the effect of decreasing meanNN, SDSD, and SDNN after PTCA was not significant for patients with prior MI. This suggest that the impact of prior MI is reflected in the modification of the intrinsic cardiac properties, such as electrical conductivity and propagation, that give place to more diverse adjustments in regulatory mechanisms that impact heart rate, including the cardiac autonomic modulation [31].

In a study where 15-min ECG recordings were used for HRV analysis, HF (n.u.) decreased after PCI, but not LF (n.u.) or LF/HF ratio, although SDNN also decreased [32], contrary to another study [28], where the HF (ms^2) increased after PTCA;nevertheless, these studies are not comparable due to the methodological differences in the measurement of HRV indices. However, in our study, regarding frequency–domain measures we found differences only in the LF and HF indices measured on power units (ms^2) but did not find any significant differences in the LF and HF indices in normalized units (n.u.) or the LF/HF ratio. The decrease in LF (ms^2) and HF (ms^2) after PTCA is more likely an effect of the overall decrease in variability (as observed in the time domain indices) which corresponds to an overall decrease in total power. In contrast, lack of significant change in the proportional contributions of each frequency band of interest (LF and HF in normalized units) indicates that, on average, PTCA had no effect on the autonomic cardiac modulation [24].

Nonlinear measures are of clinical interest, as some have been proposed as predictors of mortality in AMI (detrended fluctuation analysis) [33,34]. Complexity in HRV dynamics of patients after AMI have been assessed with sophisticated methods, such as normalized complexity index, information storage, and Gaussian linear contrast, and showed that the detection of nonlinearity by surrogate data testing depends on the nonlinear property that is measured by a particular metric [8]. In the study mentioned above, the percentage of nonlinear HRV time series after AMI (in rest), using normalized complexity index, was approximately 25%, a similar value as in the present study (DET 30.9%, LAM 23.5%, after PTCA). In our study, we found evidence of nonlinearity with a robust algorithm for surrogate data testing with both nonlinear measures, DET and LAM. However, we did not find any differences in the proportion of nonlinear time series after the intervention. It is possible that the short period of time in which the measures were made was not enough for the manifestation of different dynamics in HRV caused by reperfusion therapy, as it has been suggested in a previous work, in which detrended fluctuation analysis was assessed [35].

However, the algorithm in our work preserves nonstationary behavior in surrogates in contrast with IAAFT. Thus, the time series identified as nonlinear in [8] could mean the original time series were nonstationary or nonlinear [9]. By preserving nonlinearity on surrogate data, we reinforce that the identification of nonlinearity does detect only nonlinearity [9,10,36], as is the case of our work.

The decrease in meanNN and SDSD, and the increase in DET and LAM immediately after PTCA suggest that ischemia-reperfusion does have an effect on the HRV dynamics, in which the autonomic response on cardiac tissue may play a role [37], but it is likely that the biological changes within the cardiac tissue, mainly the induction of oxidative stress, are largely responsible for the quick changes in HRV within the first 5 min after reperfusion [38]. However, more biological research is needed to establish a direct link between the effect of ischemia-reperfusion on the cardiac tissue and HRV measures, as well as the potential clinical applications that it may have.

Increased DET implies a more significant influence of previous heartbeats on the subsequent one, that is, a reduced set of variations for the dynamical states that can acquire the system. On the other hand, LAM is related to a longer permanence of the system in a particular state with the quick capability of a sudden change. These adjustments can reflect the cardiovascular system's reorganization after the PTCA procedure. Furthermore, we observed that despite the discrepancy of change in DET and LAM in some patients (Figure 7), the average increase in these nonlinear HRV indices in response to PTCA was significant regardless of the history of prior MI, which suggest that the PTCA effect of reducing the set of variations for the dynamical states in the cardiovascular system is strong enough to occur even in those patients with prior MI. Interestingly, other works have attempted to study beat-to-beat nonlinear coupling between NN intervals and arterial pressure in patients after AMI [39] and after sympathetic and parasympathetic pharmacological blockade [40], in which the physiological context might be different from acute myocardial ischemia. Wavelet-based surrogate data might be useful to test the performance of the nonlinear coupling approach.

After PTCA, the statistical variability of time series diminished (SDNN and SDSD) and the values of nonlinear measures increased (DET and LAM), although the proportion of nonlinear time series remains similar to baseline conditions. Hence neither variability in time series nor the value of nonlinear statistics is necessarily related to the "lose" or "gain" of nonlinear behavior.

Limitations

Our study is limited to patients with pathologic conditions and does not have a valid set of comparable healthy subjects to test whether myocardial ischemia is associated with less or more nonlinear time series. Moreover, it is of physiological interest to test the effect of medication and preexisting comorbidities, but the database used for this work lacks extensive clinical information that may be pertinent to link with HRV. These aspects should be considered in future studies to deepen the factors that potentially determine the nonlinear dynamics of HRV. It has been noted in other works that the proportion of HRV nonlinear time series varies when different nonlinear discriminant statistics are used [8,41], as well as the length of the time series may have an important role in nonlinearity detection, these issues are yet to be studied in future works.

5. Conclusions

We confirmed through surrogate data testing and RQA analysis that nonlinear dynamics are present in HRV time series of patients with MI in up to 29.4% of them. The proportion of nonlinear time series did not change significantly immediately after PTCA. After the revascularization procedure, determinism and laminarity increased, which may reflect a different arrangement in data, i.e., there is a change in the dynamic behavior of HRV. However, there was no change in the presence of nonlinear dynamics, despite the loss of statistical variance. Therefore, the HRV dynamic behavior relying on traditional and nonlinear RQA measures should be interpreted with caution.

Author Contributions: Conceptualization, M.C.-J. and C.L.; methodology, M.C.-J. and C.L.; software, M.C.-J. and C.L.; formal analysis, M.C.-J., I.B.C.-V. and C.L.; investigation, M.C-J.; resources, C.L.; data curation, M.C.-J., I.B.C.-V. and C.L.; writing—original draft preparation, M.C.-J., I.B.C.-V., G.H.G.-G. and C.L.; writing—review and editing, M.C.-J., I.B.C.-V., G.H.G.-G. and C.L.; visualization, M.C.-J.,

I.B.C.-V. and C.L.; supervision, C.L.; project administration, C.L.; funding acquisition, M.C.-J. and C.L. All authors have read and agreed to the published version of the manuscript.

Funding: This research was funded by the Consejo Nacional de Ciencia y Tecnología (CONACYT), grant number 320858.

Institutional Review Board Statement: The study was conducted in accordance with the Declaration of Helsinki, and it was approved by the Institutional Review Board of the Instituto Nacional de Cardiología Ignacio Chávez (protocol code 22–1309, 22 April 2022).

Data Availability Statement: The data used in this study is publicly available at https://physionet.org/content/staffiii/1.0.0/ (accessed on 29 January 2023).

Acknowledgments: The authors thank Guering Eid-Lidt for advice on clinical definitions and Patricia Ledezma-López for technical support.

Conflicts of Interest: The authors declare no conflict of interest.

References

1. Huikuri, H.V.; Stein, P.K. Clinical application of heart rate variability after acute myocardial infarction. *Front. Physiol.* **2012**, *3*, 41. [CrossRef]
2. Stein, P.K.; Domitrovich, P.P.; Huikuri, H.V.; Kleiger, R.E. Traditional and nonlinear heart rate variability are each independently associated with mortality after myocardial infarction. *J. Cardiovasc. Electrophysiol.* **2005**, *16*, 13–20. [CrossRef] [PubMed]
3. Sassi, R.; Cerutti, S.; Lombardi, F.; Malik, M.; Huikuri, H.V.; Peng, C.-K.; Schmidt, G.; Yamamoto, Y.; Document, R.; Gorenek, B.; et al. Advances in heart rate variability signal analysis: Joint position statement by the e-Cardiology ESC Working Group and the European Heart Rhythm Association co-endorsed by the Asia Pacific Heart Rhythm Society. *EP Eur.* **2015**, *17*, 1341–1353. [CrossRef]
4. Perkiömäki, J.S. Heart rate variability and non-linear dynamics in risk stratification. *Front. Physiol.* **2011**, *2*, 81. [CrossRef]
5. Wessel, N.; Malberg, H.; Bauernschmitt, R.; Kurths, J. Nonlinear Methods of Cardiovascular Physics and Their Clinical Applicability. *Int. J. Bifurc. Chaos* **2007**, *17*, 3325–3371. [CrossRef]
6. Schreiber, T.; Schmitz, A. Surrogate time series. *Phys. D Nonlinear Phenom.* **2000**, *142*, 346–382. [CrossRef]
7. Vaillancourt, D.E.; Newell, K.M. Complexity in aging and disease: Response to commentaries. *Neurobiol. Aging* **2002**, *23*, 27–29. [CrossRef]
8. Faes, L.; Gómez-Extremera, M.; Pernice, R.; Carpena, P.; Nollo, G.; Porta, A.; Bernaola-Galván, P. Comparison of methods for the assessment of nonlinearity in short-term heart rate variability under different physiopathological states. *Chaos* **2019**, *29*, 123114. [CrossRef] [PubMed]
9. Lancaster, G.; Iatsenko, D.; Pidde, A.; Ticcinelli, V.; Stefanovska, A. Surrogate data for hypothesis testing of physical systems. *Phys. Rep.* **2018**, *748*, 1–60. [CrossRef]
10. Keylock, C.J. A wavelet-based method for surrogate data generation. *Phys. D Nonlinear Phenom.* **2007**, *225*, 219–228. [CrossRef]
11. Keylock, C.J. Characterizing the structure of nonlinear systems using gradual wavelet reconstruction. *Nonlinear Process. Geophys.* **2010**, *17*, 615–632. [CrossRef]
12. Calderón-Juárez, M.; González Gómez, G.H.; Echeverría, J.C.; Pérez-Grovas, H.; Quintanar, E.; Lerma, C. Recurrence Quantitative Analysis of Wavelet-Based Surrogate Data for Nonlinearity Testing in Heart Rate Variability. *Front. Physiol.* **2022**, *13*, 23. [CrossRef] [PubMed]
13. Marwan, N.; Wessel, N.; Meyerfeldt, U.; Schirdewan, A.; Kurths, J. Recurrence-plot-based measures of complexity and their application to heart-rate-variability data. *Phys. Rev. E* **2002**, *66*, 026702. [CrossRef]
14. Benitez, R.; Alvarez-Lacalle, E.; Echebarria, B.; Gomis, P.; Vallverdu, M.; Caminal, P. Characterization of the nonlinear content of the heart rate dynamics during myocardial ischemia. *Med. Eng. Phys.* **2009**, *31*, 660–667. [CrossRef] [PubMed]
15. Magrans, R.; Gomis, P.; Caminal, P.; Wagner, G. Multifractal and nonlinear assessment of autonomous nervous system response during transient myocardial ischaemia. *Physiol. Meas.* **2010**, *31*, 565. [CrossRef]
16. Gomis, P.; Caminal, P.; Vallverdú, M.; Warren, S.G.; Stein, P.K.; Wagner, G.S. Assessment of autonomic control of the heart during transient myocardial ischemia. *J. Electrocardiol.* **2012**, *45*, 82–89. [CrossRef]
17. Szydlo, K.; Trusz-Gluza, M.; Drzewiecki, J.; Wozniak-Skowerska, I.; Szczogiel, J. Correlation of Heart Rate Variability Parameters and QT Interval in Patients After PTCA of Infarct Related Coronary Artery as an Indicator of Improved Autonomic Regulation. *Pacing Clin. Electrophysiol.* **1998**, *21*, 2407–2410. [CrossRef]
18. Wennerblom, B.; Lurje, L.; Solem, J.; Tygesen, H.; Udén, M.; Vahisalo, R.; Hjalmarson, Å. Reduced Heart Rate Variability in Ischemic Heart Disease Is Only Partially Caused by Ischemia. *Cardiology* **2000**, *94*, 146–151. [CrossRef]
19. Tsai, M.-W.; Chie, W.-C.; Kuo, T.B.; Chen, M.-F.; Liu, J.-P.; Chen, T.H.-H.; Wu, Y.-T. Effects of Exercise Training on Heart Rate Variability After Coronary Angioplasty. *Phys. Ther.* **2006**, *86*, 626–635. [CrossRef]

20. Goldberger, A.L.; Amaral, L.A.; Glass, L.; Hausdorff, J.M.; Ivanov, P.C.; Mark, R.G.; Mietus, J.E.; Moody, G.B.; Peng, C.K.; Stanley, H.E. PhysioBank, PhysioToolkit, and PhysioNet: Components of a new research resource for complex physiologic signals. *Circulation* **2000**, *101*, E215–E220. [CrossRef]
21. Martínez, J.P.; Pahlm, O.; Ringborn, M.; Warren, S.; Laguna, P.; Sörnmo, L. The STAFF III database: ECGs recorded during acutely induced myocardial ischemia. In Proceedings of the 2017 Computing in Cardiology (CinC), Rennes, France, 24–27 September 2017; pp. 1–4.
22. Wessel, N.; Voss, A.; Malberg, H.; Ziehmann, C.; Voss, H.U.; Schirdewan, A.; Meyerfeldt, U.; Kurths, J. Nonlinear analysis of complex phenomena in cardiological data. *Herzschr. Elektrophys.* **2000**, *11*, 159–173. [CrossRef]
23. Behar, J.A.; Rosenberg, A.A.; Weiser-Bitoun, I.; Shemla, O.; Alexandrovich, A.; Konyukhov, E.; Yaniv, Y. PhysioZoo: A Novel Open Access Platform for Heart Rate Variability Analysis of Mammalian Electrocardiographic Data. *Front. Physiol.* **2018**, *9*, 1390. [CrossRef]
24. Task Force of the European Society of Cardiology and the North American Society of Pacing and Electrophysiology. Heart rate variability: Standards of measurement, physiological interpretation and clinical use. *Circulation* **1996**, *93*, 1043–1065. [CrossRef]
25. Marwan, N.; Carmen Romano, M.; Thiel, M.; Kurths, J. Recurrence plots for the analysis of complex systems. *Phys. Rep.* **2007**, *438*, 237–329. [CrossRef]
26. Javorka, M.; Turianikova, Z.; Tonhajzerova, I.; Javorka, K.; Baumert, M. The effect of orthostasis on recurrence quantification analysis of heart rate and blood pressure dynamics. *Physiol. Meas.* **2009**, *30*, 29–41. [CrossRef]
27. Schreiber, T.; Schmitz, A. Improved Surrogate Data for Nonlinearity Tests. *Phys. Rev. Lett.* **1996**, *77*, 635–638. [CrossRef]
28. Bonnemeier, H.; Hartmann, F.; Wiegand, U.K.H.; Irmer, C.; Kurz, T.; Tölg, R.; Katus, H.A.; Richardt, G. Heart rate variability in patients with acute myocardial infarction undergoing primary coronary angioplasty. *Am. J. Cardiol.* **2000**, *85*, 815–820. [CrossRef]
29. Abrootan, S.; Yazdankhah, S.; Payami, B.; Alasti, M. Changes in Heart Rate Variability Parameters after Elective Percutaneous Coronary Intervention. *J. Tehran Heart Cent.* **2015**, *10*, 80–84.
30. Coviello, I.; Pinnacchio, G.; Laurito, M.; Stazi, A.; Battipaglia, I.; Barone, L.; Mollo, R.; Russo, G.; Villano, A.; Sestito, A.; et al. Prognostic role of heart rate variability in patients with ST-segment elevation acute myocardial infarction treated by primary angioplasty. *Cardiology* **2013**, *124*, 63–70. [CrossRef]
31. Shaffer, F.; McCraty, R.; Zerr, C.L. A healthy heart is not a metronome: An integrative review of the heart's anatomy and heart rate variability. *Front. Psychol.* **2014**, *5*, 1040. [CrossRef]
32. Seetharam, S.P.; Ms, V.S.; Udupa, K.; Reddy, N.; Raveesha, A. Alterations in heart rate variability before and after percutaneous coronary intervention in patients with ischaemic heart disease. *Indian J. Physiol. Pharmacol.* **2022**, *66*, 188–195. [CrossRef]
33. Huikuri, H.V.; Mäkikallio, T.H.; Peng, C.-K.; Goldberger, A.L.; Hintze, U.; Møller, M. Fractal Correlation Properties of R-R Interval Dynamics and Mortality in Patients with Depressed Left Ventricular Function After an Acute Myocardial Infarction. *Circulation* **2000**, *101*, 47–53. [CrossRef] [PubMed]
34. Tapanainen, J.M.; Thomsen, P.E.; Køber, L.; Torp-Pedersen, C.; Mäkikallio, T.H.; Still, A.M.; Lindgren, K.S.; Huikuri, H.V. Fractal analysis of heart rate variability and mortality after an acute myocardial infarction. *Am. J. Cardiol.* **2002**, *90*, 347–352. [CrossRef]
35. Lerma, C.; Echeverria, J.C.; Infante, O.; Perez-Grovas, H.; Gonzalez-Gomez, H. Sign and magnitude scaling properties of heart rate variability in patients with end-stage renal failure: Are these properties useful to identify pathophysiological adaptations? *Chaos* **2017**, *27*, 093906. [CrossRef]
36. Faes, L.; Zhao, H.; Chon, K.H.; Nollo, G. Time-varying surrogate data to assess nonlinearity in nonstationary time series: Application to heart rate variability. *IEEE Trans. Biomed. Eng.* **2009**, *56*, 685–695. [CrossRef] [PubMed]
37. Vaseghi, M.; Shivkumar, K. The role of the autonomic nervous system in sudden cardiac death. *Prog. Cardiovasc. Dis.* **2008**, *50*, 404–419. [CrossRef]
38. Wu, M.Y.; Yiang, G.T.; Liao, W.T.; Tsai, A.P.; Cheng, Y.L.; Cheng, P.W.; Li, C.Y.; Li, C.J. Current Mechanistic Concepts in Ischemia and Reperfusion Injury. *Cell Physiol. Biochem.* **2018**, *46*, 1650–1667. [CrossRef]
39. Nollo, G.; Faes, L.; Pellegrini, B.; Porta, A.; Antolini, R. Synchronization index for quantifying nonlinear causal coupling between RR interval and systolic arterial pressure after myocardial infarction. In Proceedings of the Computers in Cardiology 2000 Conference, Cambridge, MA, USA, 24–27 September 2000; Cat. 00CH37163. Volume 27, pp. 143–146.
40. Bai, Y.; Siu, K.L.; Ashraf, S.; Faes, L.; Nollo, G.; Chon, K.H. Nonlinear coupling is absent in acute myocardial patients but not healthy subjects. *Am. J. Physiol. Heart Circ. Physiol.* **2008**, *295*, H578–H586. [CrossRef] [PubMed]
41. Porta, A.; Bari, V.; Marchi, A.; De Maria, B.; Cysarz, D.; Van Leeuwen, P.; Takahashi, A.C.M.; Catai, A.M.; Gnecchi-Ruscone, T. Complexity analyses show two distinct types of nonlinear dynamics in short heart period variability recordings. *Front. Physiol.* **2015**, *6*, 71. [CrossRef]

entropy

MDPI

Article

On the Different Abilities of Cross-Sample Entropy and K-Nearest-Neighbor Cross-Unpredictability in Assessing Dynamic Cardiorespiratory and Cerebrovascular Interactions

Alberto Porta [1,2,*], Vlasta Bari [1,2], Francesca Gelpi [1], Beatrice Cairo [1], Beatrice De Maria [3], Davide Tonon [4], Gianluca Rossato [4] and Luca Faes [5]

1 Department of Biomedical Sciences for Health, University of Milan, 20133 Milan, Italy; vlasta.bari@unimi.it (V.B.); francesca.gelpi@unimi.it (F.G.); beatrice.cairo@unimi.it (B.C.)
2 Department of Cardiothoracic, Vascular Anesthesia and Intensive Care, IRCCS Policlinico San Donato, 20097 Milan, Italy
3 IRCCS Istituti Clinici Scientifici Maugeri, 20138 Milan, Italy; beatrice.demaria@icsmaugeri.it
4 Department of Neurology, IRCCS Sacro Cuore Don Calabria Hospital, 37024 Verona, Italy; davide.tonon@sacrocuore.it (D.T.); gianluca.rossato@sacrocuore.it (G.R.)
5 Department of Engineering, University of Palermo, 90128 Palermo, Italy; luca.faes@unipa.it
* Correspondence: alberto.porta@unimi.it; Tel.: +39-02-52774382

Abstract: Nonlinear markers of coupling strength are often utilized to typify cardiorespiratory and cerebrovascular regulations. The computation of these indices requires techniques describing nonlinear interactions between respiration (R) and heart period (HP) and between mean arterial pressure (MAP) and mean cerebral blood velocity (MCBv). We compared two model-free methods for the assessment of dynamic HP–R and MCBv–MAP interactions, namely the cross-sample entropy (CSampEn) and k-nearest-neighbor cross-unpredictability (KNNCUP). Comparison was carried out first over simulations generated by linear and nonlinear unidirectional causal, bidirectional linear causal, and lag-zero linear noncausal models, and then over experimental data acquired from 19 subjects at supine rest during spontaneous breathing and controlled respiration at 10, 15, and 20 breaths·minute^{-1} as well as from 13 subjects at supine rest and during 60° head-up tilt. Linear markers were computed for comparison. We found that: (i) over simulations, CSampEn and KNNCUP exhibit different abilities in evaluating coupling strength; (ii) KNNCUP is more reliable than CSampEn when interactions occur according to a causal structure, while performances are similar in noncausal models; (iii) in healthy subjects, KNNCUP is more powerful in characterizing cardiorespiratory and cerebrovascular variability interactions than CSampEn and linear markers. We recommend KNNCUP for quantifying cardiorespiratory and cerebrovascular coupling.

Keywords: model-free time series analysis; causality; coupling strength; cardiac control; cerebral autoregulation; heart rate variability; blood flow; arterial pressure; autonomic nervous system; controlled breathing; head-up tilt

1. Introduction

There are an increasing number of studies assessing the degree of coupling between respiration (R) and heart period (HP) [1–7] and between mean arterial pressure (MAP) and mean cerebral blood velocity (MCBv) [8–14]. This interest is justified by clinical relevance: indeed, the degree of HP–R coupling is taken as a marker of vagal control being inherently normalized by breathing activity [1–7], while the strength of the MCBv–MAP relationship is an indicator of the efficiency of dynamic cerebral autoregulation (dCA) [8–14].

The dynamic interactions between R and HP and between MAP and MCBv feature common characteristics: (i) they are inherently nonlinear; (ii) their strength is between the maximum and the minimum values found in the presence of full coupling and perfect uncoupling, respectively.

Nonlinearities between R and HP result from the periodical inhibition of aortic arch, carotid sinus, pulmonary, and atrial stretch receptor activity with R [15–17], reciprocal influences between sympathetic and vagal activities [18–20], variations of HP variability with both breathing rate and depth [21–23], phase locking between heartbeat and R rhythm [24–26], and the gating activity of respiratory centers over vagal and sympathetic outflows [27–30]. On the other hand, nonlinearities between MAP and MCBv are the result of the shape of the static characteristic of cerebrovascular autoregulation [31], different MCBv responses to positive and negative MAP variations [32,33], respiratory modulations of the MCBv–MAP relationship [34], effects of intracranial pressure on the critical closing pressure [35], and nonlinear influences of the autonomic control, especially over long time scales [36].

The imperfect association between R and HP is the result of the huge number of control mechanisms adapting HP regardless of R [37,38] and the variable cardiorespiratory phase locking ratio [24,25,39]. The imperfect association between MAP and MCBv is the consequence of the dCA aiming at keeping MCBv constant by buffering MAP changes with suitable adaptations to vessel diameter [8–11].

Given the abovementioned features of the HP–R and MCBv–MAP couplings, nonlinear tools should be preferred for the assessment of the strength of the interactions and these methods should be reliable in the presence of weak relationships. The reliability of these tools is a critical issue because modifications to HP–R and MCBv–MAP coupling strength are hallmarks of pathology [1,2,8,9].

Cross-sample entropy (CSampEn) [40] and k-nearest-neighbor cross-unpredictability (KNNCUP) [41] are two methods devised to assess nonlinear interactions between two time series. The ability of CSampEn and KNNCUP to describe nonlinear dynamics lies in their model-free nature that does not impose any form to the underlying relationship between the observed series. After reconstructing the dynamics of the two series in two distinct state spaces built via the time-delay embedding procedure, CSampEn estimates the negative logarithm of the conditional probability that, if two patterns of length $m-1$ are similar, they remain alike after including one additional future value: the higher the conditional probability, the smaller the negative logarithm, the stronger the relationship and the coupling between the two series [40]. KNNCUP exploits the k patterns built over $m-1$ past samples of the driver signal, namely R and MAP in our application, selected among others for their similarity with the reference vector, to predict the current value of the target signal, namely HP and MCBv, respectively: the lower the unpredictability of the target using the driver, the stronger the coupling from the driver to the target [41].

The aim of the present study is to compare the abilities of CSampEn and KNNCUP to characterize cardiorespiratory and cerebrovascular couplings via the analysis of the variability of R and HP and of MAP and MCBv. The two tools were first tested over simulations generated by linear and nonlinear unidirectional causal, linear bidirectional causal, and lag-zero linear noncausal models to better understand eventual differences in assessing dynamic HP–R and MCBv–MAP interactions. CSampEn and KNNCUP were then assessed in healthy subjects under experimental protocols inducing modifications of the cardiorespiratory and cerebrovascular controls, namely controlled breathing (CB) at different breathing rates [42,43] and head-up tilt (HUT) with a tilt table inclination of 60° [44,45]. Results of CSampEn and KNNCUP were compared with linear markers of coupling [4,9]. Preliminary data were presented at the 44th Annual International Conference of the Engineering in Medicine and Biology Society [46].

2. Methods

2.1. Generalities for the Computation of CSampEn and KNNCUP

Given two systems X and Y, their joint activity is described by two realizations, $x = \{x_n, 1 \le n \le N\}$ and $y = \{y_n, 1 \le n \le N\}$, of the stochastic processes X and Y, respectively. The series x and y are usually collections of values recorded in experimental sessions designed to probe into the interactions between X and Y. We denote: (i) the current value of y as y_i; (ii) the pattern formed by the $m-1$ past values of y_i as $\mathbf{y}_i^- = [y_{i-1} \ \cdots \ y_{i-m+1}]$; and (iii) the m-

dimensional vector obtained by concatenating y_i with $\boldsymbol{y}_i^-$ as $\boldsymbol{y}_i = \begin{bmatrix} y_i & y_{i-1} & \cdots & y_{i-m+1} \end{bmatrix}$. We remark that $\boldsymbol{y}_i^-$ and $\boldsymbol{y}_i$ can be interpreted indifferently as patterns of length, respectively, and $m - 1$ and m in the time domain or points of, respectively, the $(m - 1)$- and m-dimensional state phase spaces built uniformly using the method of lagged coordinates in the state phase domain. Analogously, we define x_j, $\boldsymbol{x}_j^- = \begin{bmatrix} x_{j-1} & \cdots & x_{j-m+1} \end{bmatrix}$, and $\boldsymbol{x}_j = \begin{bmatrix} x_j & x_{j-1} & \cdots & x_{j-m+1} \end{bmatrix}$ the equivalent quantities computed over x. In order to characterize the relationship, if present, between X and Y, we indicate the probability that $\boldsymbol{y}_i$ and $\boldsymbol{x}_j$ are close in the m-dimensional state phase space within a tolerance r with $p(\|\boldsymbol{y}_i - \boldsymbol{x}_j\| \leq r)$ and with $p\left(\left\|\boldsymbol{y}_i^- - \boldsymbol{x}_j^-\right\| \leq r\right)$ the probability that $\boldsymbol{y}_i^-$ and $\boldsymbol{x}_j^-$ are closer than r in the $(m - 1)$-dimensional state phase space, where $\|\cdot\|$ is a metric to compute distance. In this study the adopted metric is the Euclidean norm.

2.2. CSampEn

CSampEn [40] is defined as the negative logarithm of the ratio of the averaged $p(\|\boldsymbol{y}_i - \boldsymbol{x}_j\| \leq r)$ to the averaged $p\left(\left\|\boldsymbol{y}_i^- - \boldsymbol{x}_j^-\right\| \leq r\right)$ as

$$\mathrm{CSampEn}(m,\ r,\ N) = -\log\left(\frac{\langle p(\|\boldsymbol{y}_i - \boldsymbol{x}_j\| \leq r)\rangle}{\left\langle p\left(\left\|\boldsymbol{y}_i^- - \boldsymbol{x}_j^-\right\| \leq r\right)\right\rangle}\right), \tag{1}$$

where $\langle\cdot\rangle$ performs the time average over all the reference vectors built from x and $\log(\cdot)$ is the natural logarithm. $p(\|\boldsymbol{y}_i - \boldsymbol{x}_j\| \leq r)$ and $p\left(\left\|\boldsymbol{y}_i^- - \boldsymbol{x}_j^-\right\| \leq r\right)$ are estimated by counting the number of $\boldsymbol{y}_i$ closer than r to $\boldsymbol{x}_j$ for $i = m, \ldots, N$ and the number of $\boldsymbol{y}_i^-$ closer than r to $\boldsymbol{x}_j^-$ for $i = m - 1, \ldots, N - 1$ and by dividing them by $N - m + 1$. The unfortunate case was that $\langle p(\|\boldsymbol{y}_i - \boldsymbol{x}_j\| \leq r)\rangle = 0$ and/or $\left\langle p\left(\left\|\boldsymbol{y}_i^- - \boldsymbol{x}_j^-\right\| \leq r\right)\right\rangle = 0$ were never observed in our study given the adopted values of m and N. CSampEn is a measure of the negative logarithm of the conditional probability that two points remain close in the m-dimensional phase space given that they are close in the $(m - 1)$-dimensional phase space: the closer to 1 the conditional probability, the lower the uncertainty associated with the reciprocal position of patterns in the m-dimensional phase space, the smaller the CSampEn, the stronger the relationship between X and Y.

2.3. KNNCUP

Cross-unpredictability (CUP) searches for the dependency $f(\cdot)$ of the future value $y_{i+\tau}$ of the dynamics of the target Y on $m - 1$ past samples $\boldsymbol{x}_i^-$ of the dynamics of the driver X [41,47], where τ is the prediction horizon [14]. $y_{i+\tau}$ is usually labelled as the image of $\boldsymbol{x}_i^-$ through $f(\cdot)$ and $\boldsymbol{x}_i^-$ is the reference vector for the search of its k nearest neighbors. Local CUP approach approximates $f(\cdot)$ in a region around $\boldsymbol{x}_i^-$ set by its k nearest neighbors [48]. Given the adopted metric for evaluating distances in the phase space, the region is a hypersphere and $\boldsymbol{x}_i^-$ is its center. The k nearest neighbors of the reference vector $\boldsymbol{x}_i^-$ were utilized to predict $y_{i+\tau}$. The prediction of $y_{i+\tau}$, namely $\hat{y}_{i+\tau}$, is defined as the weighted mean of the images of the k nearest neighbors of $\boldsymbol{x}_i^-$, where the weights are the inverse of their distance from the reference vector. Vectors at zero distance from $\boldsymbol{x}_i^-$ were excluded from the set of k nearest neighbors. $\hat{y}_{i+\tau}$ was computed as a function of time i, thus providing the predicted series $\hat{y}$. The degree of unpredictability of Y given X is measured via the CUP function defined as $1 - \rho^2$, where ρ^2 is the squared correlation coefficient between y and $\hat{y}$. CUP is bounded between 0 and 1, where 0 indicates that y can be perfectly predicted from x and 1 indicates complete unpredictability of y based on x. CUP depends on m. The course of CUP with m was the result of two opposite tendencies [49]: (i) prediction of y given x might improve and CUP might decrease because longer patterns built over x bring more information about the future behavior of y; (ii) at high m, the k nearest neighbors tend to be far away from the reference vector due to the spreading of the vectors in the phase space and this effect raises CUP. The minimum of the CUP function over m was taken as a measure of the uncoupling between x and y [41] and referred to as the CUP index (CUPI): the closer to 0 the CUPI, the greater the ability to predict y based on x, the stronger the relationship between X and Y.

3. Simulations

3.1. Graded Unidirectional and Bidirectional Causal Couplings

We simulated bivariate autoregressive (BAR) processes whose components feature different degrees of coupling and causal relationships. The BAR process is defined as

$$\begin{aligned} X(n) &= 2\rho_1 \cdot [c_1 \cdot Y(n-1) + (1-c_1) \cdot X(n-1)] \cdot \cos\varphi_1 - \rho_1^2 \cdot X(n-2) + W_1(n) \\ Y(n) &= 2\rho_2 \cdot [c_2 \cdot X(n-1) + (1-c_2) \cdot Y(n-1)] \cdot \cos\varphi_2 - \rho_2^2 \cdot Y(n-2) + W_2(n) \end{aligned} \quad (2)$$

where W_1 and W_2 are Gaussian white noises with zero mean and variances assigned such that X and Y exhibit unit variance. If $c_1 = 0$ and $c_2 = 0$, X and Y are two uncoupled second-order autoregressive [AR(2)] processes. If $c_1 = 0$ and $c_2 \neq 0$, the directionality of the interactions is from X to Y (i.e., unidirectional causal model). If $c_1 \neq 0$ and $c_2 \neq 0$, the directionality of the interactions is from X to Y and vice versa (i.e., bidirectional causal model). In the uncoupled condition, the two processes X to Y were set to feature dominant rhythms according to two configurations: (i) $\rho_1 = \rho_2 = 0.8$ with phases $\varphi_1 = \pm 3\pi/5$ and $\varphi_2 = \pm 3\pi/5$ corresponding to an oscillation at normalized frequency $f_1 = 0.3$ cycles·sample^{-1} and $f_2 = 0.3$ cycles·sample^{-1}, thus simulating a dominant rhythm in the high frequency (HF) band typical of spontaneous respiration driving an HP rhythm at the same frequency with a mean HP equal to 1 s; (ii) $\rho_1 = \rho_2 = 0.8$ with phases $\varphi_1 = \pm\pi/5$ and $\varphi_2 = \pm\pi/5$ corresponding to an oscillation at normalized frequency $f_1 = 0.1$ cycles·sample^{-1} and $f_2 = 0.1$ cycles·sample^{-1}, thus leading to a dominant rhythm in the low frequency (LF) band typical of slow breathing driving an HP rhythm at the same frequency with a mean HP equal to 1 s. Coupling strengths between X and Y were varied according to the following setup: (i) $c_1 = 0$ and c_2 was varied incrementally from 0 to 1.0 in 0.1 steps, thus simulating unidirectional coupling from X to Y with incremental coupling strength (i.e., from the respiratory system to the heart or from the systemic to cerebral vasculature); $c_1 = c_2 = c$, and c was varied gradually from 0 to 1.0 in 0.1 steps, thus simulating bidirectional coupling from X to Y and vice versa with incremental coupling strength (i.e., from the respiratory system to the heart, or from the systemic to cerebral vasculature, and vice versa).

3.2. Graded Lag-Zero Noncausal Coupling

We simulated a bivariate process whose components featured different degrees of coupling in absence of any causal relationship. This bivariate process is obtained by corrupting an AR(2) process X with a Gaussian white noise W_2 with zero mean and standard deviation taken as a fraction of the standard deviation of X. The bivariate process is defined as

$$Y(n) = X(n) + W_2(n), \quad (3)$$

where X is an AR(2) process with zero mean and unit standard deviation and W_2 is a Gaussian white noise with zero mean and standard deviation $1 - c_2$. As in Section 3.1, X exhibits dominant rhythms according to two pole configurations: (i) $\rho_1 = \rho_2 = 0.8$ with phases $\varphi_1 = \pm 3\pi/5$ and $\varphi_2 = \pm 3\pi/5$; (ii) $\rho_1 = \rho_2 = 0.8$ with phases $\varphi_1 = \pm\pi/5$ and $\varphi_2 = \pm\pi/5$. Since interactions between X and Y occur at lag-zero, directionality is not set (i.e., the coupling is instantaneous), thus simulating noncausal interactions between X and Y. If $c_2 = 1$, X and Y are coincident and, assigned the outcome of the random experiment, the points $[x(n), y(n)]$ lie exactly on the diagonal line for $n = 1, \ldots, N$, and the two processes X and Y are fully coupled. c_2 was varied incrementally from 0 to 1.0 in 0.1 steps, thus progressively decreasing the standard deviation of W_2 and increasing the coupling strength between X and Y.

3.3. Unidirectionally-Coupled Identical Logistic Maps

Nonlinear dynamics were simulated via logistic maps. We considered two unidirectionally-coupled identical logistic maps [50], described as

$$\begin{aligned} X(n) &= f[X(n-1)] \\ Y(n) &= c_2 \cdot f[X(n-1)] + (1-c_2) \cdot f[Y(n-1)], \end{aligned} \tag{4}$$

with

$$\begin{aligned} f[X(n-1)] &= r \cdot X(n-1) \cdot [1 - X(n-1)] \\ f[Y(n-1)] &= r \cdot Y(n-1) \cdot [1 - Y(n-1)]. \end{aligned} \tag{5}$$

Thus, X and Y are deterministic signals obtained by iterating (4) starting from the initial conditions randomly chosen within the interval from 0.0 to 1.0. When $c_2 = 0$, X and Y evolve independently of each other because their initial conditions are different according to the equations of the two logistic maps. Given that $r = 3.7$, the two logistic maps are in chaotic regime. The coupling is unidirectional because Y does not affect X. The strength of the interactions from X to Y increases with c_2. The parameter c_2 was varied from 0 to 1.0 in 0.1 steps.

4. Experimental Protocol and Data Analysis

4.1. Ethical Statement

The R and HP series were extracted from a historical database built to evaluate changes in the cardiorespiratory coupling with the breathing rate in healthy subjects [42,43]. The MAP and MCBv series belonged to another historical database built to study the dCA during postural stimulus in healthy subjects [44,45]. All the original protocols were in keeping with the Declaration of Helsinki. The protocols were approved by the local ethical review board of the L. Sacco Hospital, Milan, Italy, and Sacro Cuore Don Calabria Hospital, Negrar, Italy, respectively, and authorized by the same structures. Written signed informed consent was obtained from all subjects. Physical examination and full neurological evaluation certified the healthy status of all the subjects. The subjects were not under pharmacological treatments interfering with cardiovascular and cerebrovascular controls. Subjects avoided caffeinated and alcoholic beverages and heavy physical exercises for 24 h before the study.

4.2. CB Protocol

Data were acquired from 19 healthy subjects (age: 27–35 years, median = 31 years; 8 males, 11 females) at rest in a supine position during spontaneous breathing (SB) and during CB at 10, 15, and 20 breaths·minute^{-1}, labelled as CB10, CB15, and CB20, respectively. The period of SB always preceded the session of CB. The respiratory frequency of the CB was selected randomly. The subjects performed all the CB sessions. The timing of inspiratory and expiratory onsets was provided via a metronome and reinforced verbally by the experimenter. All the experimental sessions lasted 10 min. The subjects were not allowed to talk during the entire protocol. The electrocardiogram (ECG) was acquired from lead II via a bioamplifier (Marazza, Monza, Italy) and respiratory flow via a nasal thermistor (Marazza, Monza, Italy). Both signals were digitalized synchronously at 300 Hz by an analog-to-digital 12-bit board (National Instruments, Austin, TX, USA) plugged into a personal computer. From the ECG, we derived the beat-to-beat variability series of the HP and a downsampled version of the R signal. After identifying the R-wave from the ECG, the inter-heartbeat interval between the nth and $(n + 1)$th R-wave peaks was taken as the nth HP. The R signal was sampled at the nth R-wave peak.

4.3. HUT Protocol

Data were acquired from 13 healthy subjects with no history of postural syncope (age: 27 ± 8 years; 5 males, 8 females). Subjects were instrumented to continuously monitor the ECG from lead II and noninvasive continuous arterial pressure (AP) from the middle finger of the nondominant arm (Finapres Medical Systems, Enschede, The Netherlands). The cerebral blood velocity (CBv), measured from the right, or left, middle cerebral artery through a transcranial Doppler device (Multi-Dop T, DWL, 2 MHz, Compumedics, San Juan Capistrano, CA, USA), was taken as a surrogate of cerebral blood flow [51]. The CBv signal was low-pass filtered with a sixth-order Butterworth filter with a cut-off frequency of 10 Hz. The signals were acquired synchronously at a sampling rate of 1000 Hz. The subjects underwent 10 min of recording at rest in a supine position (REST) followed by HUT with a tilt table inclination of 60°. Both sessions were under SB. Analyses were carried out 3 min after the HUT onset and within the first 10 min of HUT. None of the subjects exhibited presyncope signs. We utilized the detection of the R-wave on the ECG to trigger the process of identification of systolic AP (SAP) and diastolic AP (DAP). The nth SAP was taken as the maximum of the AP signal within the nth HP. The $(n-1)$th DAP was defined as the minimum of the AP signal preceding the nth SAP. The nth MAP was computed as the ratio of the definite integral of AP between the occurrences of the $(n-1)$th and nth DAP to the inter-diastolic interval. The same procedure was applied to CBv to calculate the MCBv. The fiducial points for the computation of MAP were utilized for the computation of the MCBv.

4.4. Time Domain Analysis

The HP, SAP, DAP, MAP, and MCBv series were manually checked and corrected in case of missing beats or misdetections. The effects of ectopic beats or isolated arrhythmic events were mitigated via linear interpolation. Synchronous sequences were randomly selected within the whole recordings. The length of the series was kept constant regardless of protocol and experimental condition. As to the CB protocol, we monitored the mean and variance of the HP series. These markers were labeled as μ_{HP} and σ^2_{HP}, respectively, and expressed in ms and ms^2. In the CB protocol, μ_{HP} did not vary compared to SB, being 1010 ± 168, 989 ± 157, 1023 ± 162, and 1028 ± 162 during SB, CB10, CB15, and CB20, respectively, while σ^2_{HP} increased solely during CB10 compared to SB, being 3368 ± 2622, 4784 ± 3356, 3705 ± 3091, and 2813 ± 2158 during SB, CB10, CB15, and CB20, respectively. As to the HUT series, we monitored the mean and variance of the MAP and MCBv. These indices were denoted as μ_{MAP}, σ^2_{MAP}, μ_{MCBv}, and σ^2_{MCBv} and expressed in mmHg, mmHg^2, cm·s^{-1} and cm^2·s^{-2}. In the HUT protocol, μ_{MCBv} decreased during HUT compared to REST, being 72 ± 23 and 62 ± 21, respectively, while σ^2_{MCBv} increased from 19 ± 12 to 26 ± 16. In the HUT protocol, μ_{MAP} and σ^2_{MAP} did not vary with HUT: μ_{MAP} was 99 ± 17 and 95 ± 12 at REST and during HUT, respectively, while σ^2_{MAP} was 18 ± 21 and 19 ± 12.

4.5. Computation of a Linear Marker of Association between Time Series

Squared coherence function $K^2_{x,y}(f)$ provides an estimation of the degree of linear association between x and y as a function of the frequency f [4]. It is computed as the ratio of the square cross-spectrum modulus between x and y divided by the product of their power spectra. By definition, $K^2_{x,y}(f)$ is bounded between 0 and 1, where 0 and 1 indicate the minimum and maximum correlation between x and y. A linear marker of the strength of the cardiorespiratory coupling is commonly computed by sampling $K^2_{R,HP}(f)$ at its peak in the HF band (i.e., from 0.15 to 0.4 Hz) [4,7]. This marker is indicated as $K^2_{R,HP}(HF)$ in the following. A linear marker of the strength of the cerebrovascular coupling is routinely computed by sampling $K^2_{MAP,MCBv}(f)$ at its peak in the very LF (VLF, from 0.02 to 0.07 Hz), LF (from 0.07 to 0.15 Hz), and HF (i.e., from 0.15 to 0.4 Hz) bands [8,9]. These markers are indicated as $K^2_{MAP,MCBv}(VLF)$, $K^2_{MAP,MCBv}(LF)$, and $K^2_{MAP,MCBv}(HF)$ in the following. The superior limit of the LF band and the inferior limit of the HF were modified compared to the original definition [8] to account for possible slow respiratory rhythms [12]. $K^2_{R,HP}(f)$ and $K^2_{MAP,MCBv}(f)$ were estimated according to a bivariate AR model [7]. The model order was fixed to 10, and the coefficients of the bivariate AR model were identified via the least squares approach [7].

4.6. Computation of CSampEn and KNNCUP

Since the aim of the study was to explore the physiological mechanisms responsible for short-term control of HP and MCBv, the length N of the series was set to 256 [8,52]. The series were first linearly detrended and then normalized to have zero mean and unit variance. Cardiorespiratory coupling was assessed with x = R and y = HP, while the cerebrovascular link was evaluated using x = MAP and y = MCBv. CSampEn was computed over the normalized series with m = 3 and r = 0.2 [14]. KNNCUP was performed with k = 30 [41]. Over simulated data, KNNCUP was computed with time horizon $\tau = -1$ in the case of lag-zero noncausal model and with time horizon $\tau = 0$ in the case of unidirectional and bidirectional causal models. Over experimental data, KNNCUP was carried out with $\tau = -1$ in agreement with the fast vagal actions responsible for the respiratory sinus arrhythmia [23,53] and fast resistive component of the MCBv–MAP relationship [54,55].

4.7. Statistical Analysis

One-way repeated measures analysis of variance (Dunnett's test for multiple comparisons), or Friedman repeated measures analysis of variance on ranks if appropriate (Dunnett's test for multiple comparisons), was utilized to check the effect of CB versus SB. A paired t-test, or a Wilcoxon signed rank test when appropriate, was applied to check the effect of HUT. Statistical analysis was carried out using a commercial statistical program (Sigmaplot, v.14.0, Systat Software, Inc., Chicago, IL, USA). A $p < 0.05$ was always considered statistically significant.

5. Results

5.1. Results on Simulations

The line plot of Figure 1 shows the mean (solid line) and the confidence interval of two standard deviations about the mean (dashed lines) of CSampEn (Figure 1a,c,e) and CUPI (Figure 1b,d,f) computed over 20 realizations of X and Y. Simulations were generated via linear unidirectional causal (Figure 1a,b), linear bidirectional causal (Figure 1c,d) and lag-zero linear noncausal (Figure 1e,f) models. The simulated series featured a dominant HF rhythm. Regardless of the type of simulations, the expectation is that the coupling strength increases progressively with c_2. According to this expectation, CUPI decreased gradually with c_2, and this result held in simulations of linear unidirectional causal (Figure 1b), linear bidirectional causal (Figure 1d), and lag-zero linear noncausal (Figure 1f) couplings. Conversely, CSampEn declined gradually solely in simulations relevant to the lag-zero linear noncausal model (Figure 1e), being stable over unidirectional causal coupling (Figure 1a) and paradoxically increasing in bidirectional causal interactions (Figure 1c). Results stress the limited ability of CSampEn and the more reliable performance of CUPI.

Like Figure 1, Figure 2 shows the results of simulations generated via linear unidirectional causal (Figure 2a,b), linear bidirectional causal (Figure 2c,d) and lag-zero linear noncausal (Figure 2e,f) models. However, the results are relevant to 20 realizations of X and Y featuring a dominant LF rhythm. The results stress the greater ability of CUPI in following the increased coupling strength with c_2. Indeed, CUPI gradually decreased with c_2 regarless of the type of simulations (Figure 2b,d,f). CSampEn exhibited bad performance over causal models (Figure 2a,c) and good performance in the case of lag-zero linear noncausal interactions (Figure 2e).

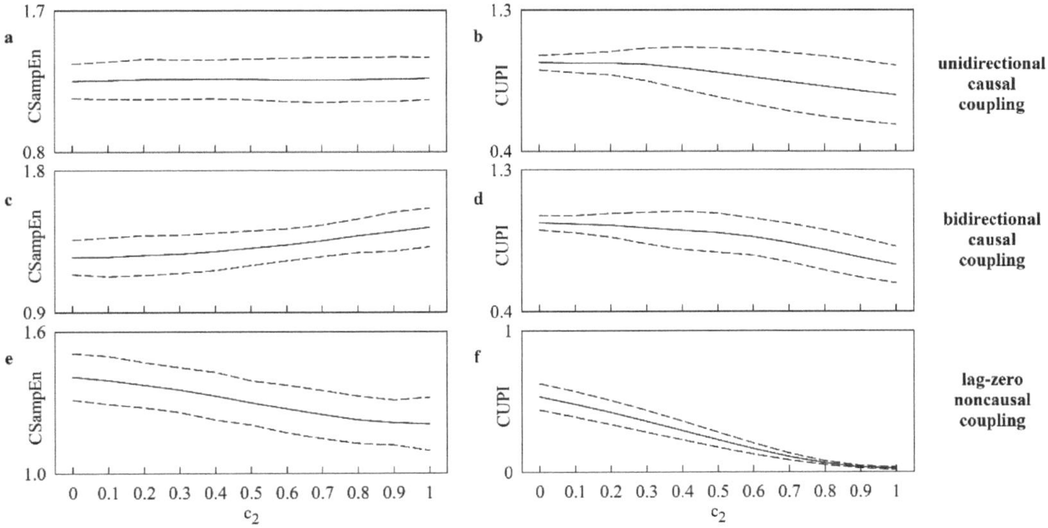

Figure 1. The line plots show the mean (solid line) and the confidence interval of two standard deviations about the mean (dashed lines) of CSampEn (**a**,**c**,**e**) and CUPI (**b**,**d**,**f**) as a function of c_2. The results of simulations generated via linear unidirectional causal (i.e., $c_1 = 0$), linear bidirectional causal (i.e., $c_1 = c_2$), and lag-zero linear noncausal models are shown in (**a**,**b**), (**c**,**d**), and (**e**,**f**), respectively. The processes exhibit a dominant HF rhythm. The curves were built over 20 realizations of X and Y.

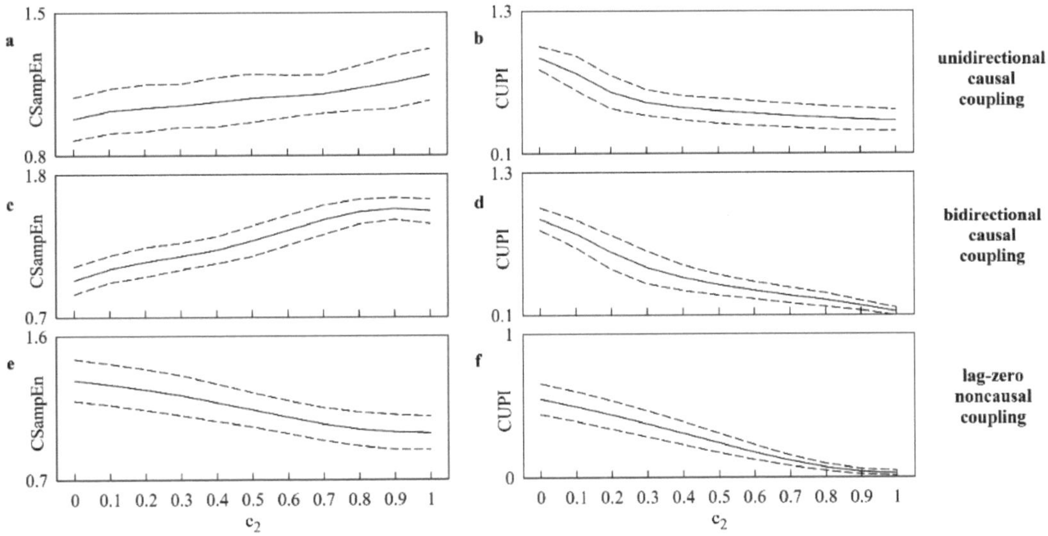

Figure 2. The line plots show the mean (solid line) and the confidence interval of two standard deviations about the mean (dashed lines) of CSampEn (**a**,**c**,**e**) and CUPI (**b**,**d**,**f**) as a function of c_2. The results of simulations generated via linear unidirectional causal (i.e., $c_1 = 0$), linear bidirectional causal (i.e., $c_1 = c_2$), and lag-zero linear noncausal models are shown in (**a**,**b**), (**c**,**d**), and (**e**,**f**), respectively. The processes exhibit a dominant LF rhythm. The curves were built over 20 realizations of X and Y.

Figure 3 shows the mean (solid line) and the confidence interval of two standard deviations about the mean (dashed lines) of CSampEn (Figure 3a) and CUPI (Figure 3b) computed over dynamics generated via unidirectionally-coupled identical logistic maps while varying c_2. The curves were built over 20 pairs of signals generated according to different initial conditions. CUPI decreased to 0 with c_2 and reached 0 when X and Y synchronized (Figure 3b). Conversely, CSampEn remained stable with c_2, and this result was the consequence of the inability of CSampEn to detect the situation of uncoupling when c_2 was close to 0 (Figure 3a).

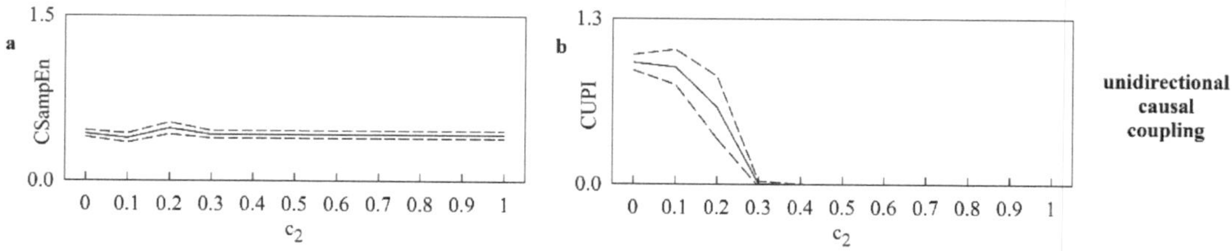

Figure 3. The line plots show the mean (solid line) and the confidence interval of two standard deviations about the mean (dashed lines) of CSampEn (**a**) and CUPI (**b**) as a function of c_2. The results are relevant to unidirectionally-coupled identical logistic maps. The curves were built over 20 pairs of X and Y generated according to different initial conditions.

5.2. Results on CB and HUT Protocols

Figure 4 shows CSampEn (Figure 4a) and CUPI (Figure 4b) computed in the CB protocol. CSampEn decreased significantly during CB10 compared to SB, thus suggesting that cardiorespiratory coupling increased at the slowest breathing rate. CUPI exhibited a similar trend, but its variation compared to SB was significant during both CB10 and CB15, thus making more evident the dependence of the cardiorespiratory coupling strength on the respiratory rate during CB.

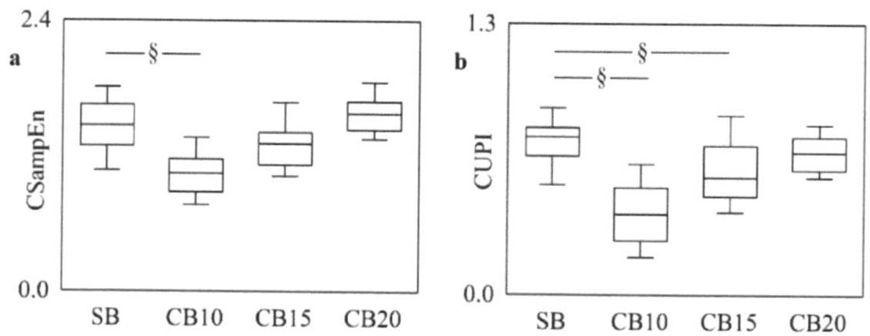

Figure 4. The vertical box-and-whisker plots show CSampEn (**a**) and CUPI (**b**) as a function of the experimental condition (i.e., SB, CB10, CB15, and CB20). The height of the box represents the distance between the first and third quartiles, with the median marked as a horizontal segment, and the whiskers denote the 5th and 95th percentiles. The symbol § indicates $p < 0.05$ versus SB.

Figure 5 shows CSampEn (Figure 5a) and CUPI (Figure 5b) computed in the HUT protocol. The two markers exhibited striking differences with the experimental condition. Indeed, CUPI decreased significantly during HUT, thus suggesting an increase of the cerebrovascular coupling strength, while CSampEn remained stable, thus suggesting a certain stiffness in following modifications of the cerebrovascular coupling with the experimental condition.

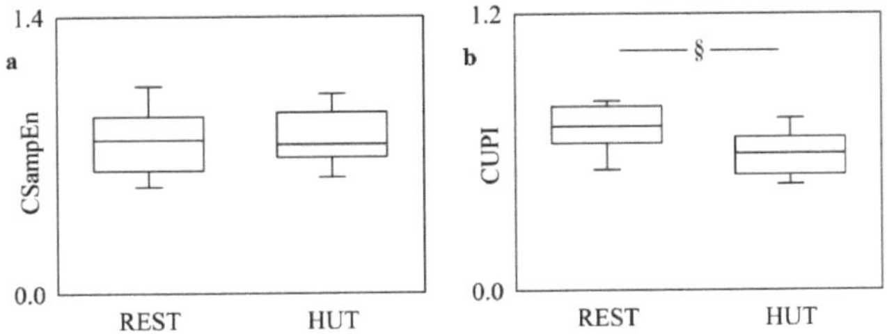

Figure 5. The vertical box-and-whisker plots show CSampEn (**a**) and CUPI (**b**) as a function of the experimental condition (i.e., REST and HUT). The height of the box represents the distance between the first and third quartiles, with the median marked as a horizontal segment, and the whiskers denote the 5th and 95th percentiles. The symbol § indicates $p < 0.05$ versus REST.

Figure 6 shows the linear markers of cardiorespiratory (Figure 6a) and cerebrovascular (Figure 6b–d) coupling calculated in the CB and HUT protocols, respectively. CB augmented $K^2_{R,HP}(HF)$ regardless of the rate of paced breathing (Figure 6a). $K^2_{MAP,MCBv}(VLF)$ and $K^2_{MAP,MCBv}(LF)$ increased during HUT, and this result indicated an increased strength of the cerebrovascular coupling (Figure 6b,c). Conversely, $K^2_{MAP,MCBv}(HF)$ did not vary during HUT (Figure 6d).

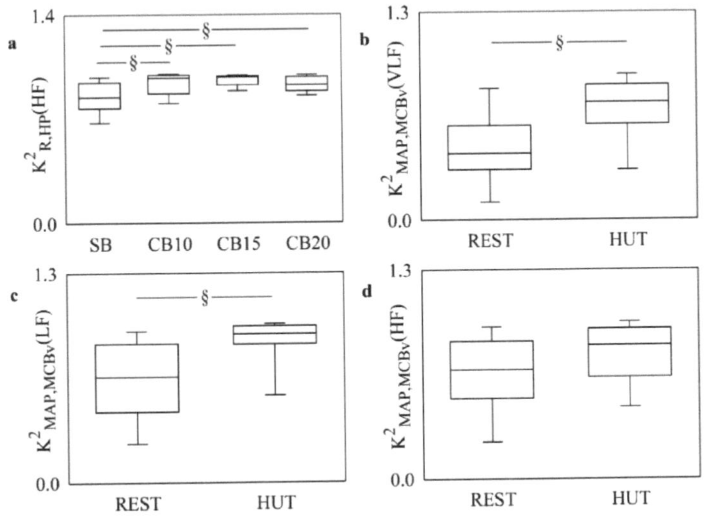

Figure 6. The vertical box-and-whisker plots show the K^2 marker computed between R and HP in the HF band in the CB protocol (**a**) and the K^2 markers computed between MAP and MCBv in the VLF (**b**), LF (**c**), and HF (**d**) bands in the HUT protocol. The height of the box represents the distance between the first and third quartiles, with the median marked as a horizontal segment, and the whiskers denote the 5th and 95th percentiles. The symbol § indicates $p < 0.05$ versus CB or REST.

6. Discussion

The main findings of this study can be summarized as follows: (i) CSampEn and KNNCUP can quantify the modifications of coupling strength but with different abilities; (ii) KNNCUP is more reliable than CSampEn when interactions occur according to a causal structure, while performances are similar over a lag-zero linear noncausal model; (iii) in healthy subjects KNNCUP is more powerful in identifying the changes in cardiorespiratory and cerebrovascular coupling strength in response to experimental challenges than CSampEn and linear markers.

6.1. Assessing the Coupling Strength between Dynamic Systems via CSampEn and KNNCUP

The degree of interaction between dynamic systems X and Y is assessed in the present study via two model-free methods, namely CSampEn and KNNCUP.

CSampEn reconstructs the dynamics of X and Y in two separate embedding spaces built using the technique of delayed components and assesses the strength of the relationship between the embedding spaces by computing the negative logarithm of the conditional probability that two patterns generated, respectively, by X and Y that are close in the $(m-1)$-dimensional embedding space remain close in the m-dimensional one. In other words, according to the philosophy of state space correspondence methods, if two patterns occupy the same region in the $(m-1)$-dimensional embedding spaces, in presence of a significant coupling between X and Y, they do not move apart when a new component is added.

Therefore, although the model-free nature of CSampEn allows for the theoretical description of nonlinear interactions, strong nonlinearities that imply links among different regions of the phase space, such as those responsible for interactions among rhythms at different frequencies, cannot be reliably described and the method seems to be more suitable for describing relations occurring according to a 1:1 coupling ratio.

However, simulations provided in this study prove that, even in the presence of a 1:1 coupling ratio typically occurring when two processes with the same dominant oscillation are considered, CSampEn cannot detect the progressive modification of the coupling strength when interactions between X and Y are generated by a linear causal model. This conclusion held even in the case of a model simulating the interactions between two nonlinear deterministic signals generated by logistic maps in a chaotic regime. This limitation is related to the inability of CSampEn to interpret causality given that it is a symmetric metric under the reversal of the role between X and Y. The most regrettable feature of CSampEn is that it can suggest even the opposite trend with the coupling strength as shown in Figures 1c and 2a,c. Solely in trivial simulations where two stochastic processes interact in absence of a causal structure (i.e., immediate interactions), CSampEn can detect the expected modifications with the coupling strength. The limited ability of CSampEn confirms that the exploitation of a metric based on conditional probability does not necessarily assure a causal approach [56].

KNNCUP assesses the relationship between X and Y using a completely different logic, namely the cross-predictability of Y from patterns taken from the past history of X. KNNCUP recalls model-based cross-conditional entropy even though the metric utilized to assess irregularity of future behaviors of Y given the activity of X is different from the evaluation of the logarithm of the variance of the innovation of Y given X [57]. Therefore, the pattern is created over the activity of X and the ability of this pattern to set future behaviors of Y is tested by measuring the ability of predicting y based on the knowledge of x. In addition to having the possibility to describe nonlinear interactions among rhythms at different frequencies according to the number of past samples of x utilized to predict y, KNNCUP could account for the causal structure of the mechanisms generating the interactions between X and Y. In addition to accounting for the directionality of the interactions, KNNCUP optimizes the embedding dimension and coarse graining via the search for the minimum over m [41,49] and the coverage of the embedding space with cells of different size according to the k-nearest-neighbor strategy [58], respectively. The most remarkable feature of KNNCUP is its robustness in indicating the expected modifications of the coupling strength regardless of the model structure (i.e., causal or noncausal), type of causal interactions (i.e., unidirectional or bidirectional), and signal feature (i.e., linear stochastic or nonlinear deterministic).

In conclusion, KNNCUP should be preferred to CSampEn as a model-free technique for the assessment of coupling strength even in the presence of interactions occurring at the same frequency.

6.2. Superior Ability of CUPI Compared to CSampEn in Evaluating Cardiorespiratory Coupling Strength

CSampEn and CUPI exhibited similar results in the CB protocol. Indeed, both methods detected an increase in cardiorespiratory coupling strength during CB10. This result is not surprising given that the same conclusion was reached via cross-conditional entropy based on a uniform quantization strategy for the computation of probabilities [42]. A linear approach based on $K^2_{R,HP}(f)$ was able to reach the same conclusion, but the trend with the breathing rate was not evident. Therefore, given the dependence of the cardiorespiratory coupling strength on the breathing rate, we recommend the use of a respiratory rate-matched control group in applications to pathological subjects. At first sight, this conclusion might be the consequence of the increase of the transfer function magnitude from R to HP while slowing the respiratory rate [4,21,22] resulting from the shape of the sinus node transfer function [59,60]. Conversely, since a rise in the transfer function magnitude is not necessarily accompanied by an increase of the coupling strength, the observed finding is more likely to be the consequence of a firmer pacing of the neural efferent activity operated by the respiratory centers [27–30]. Given that the influence of R on HP is very rapid [4,23,53] and produces dominant HP oscillations at the same frequency [22,23], the difficulties of CSampEn in assessing coupling strength are less evident in the CB protocol. However, CSampEn and CUPI are not fully equivalent. As a matter of fact, CSampEn exhibited a slightly lower statistical power in detecting the dependence of the cardiorespiratory coupling strength on the respiratory rate than CUPI. This finding suggests that some difficulties of SampEn might be linked to the presence of slower components in the responses of HP to R [5,61] and to the significant action in the reverse causal direction (i.e., from HP to R) [3,26,53]. In addition, since the likelihood of the cross-frequency coupling, for example, between the main respiratory rhythm and oscillations generated by baroreflex control loop [62,63] increases while slowing the breathing rate, the resulting impairment of the state space correspondence might have, more importantly, reduced the effectiveness of CSampEn compared to that of CUPI.

6.3. Superior Ability of CUPI Compared to CSampEn in Evaluating Cerebrovascular Coupling Strength

CUPI declined during HUT, thus indicating that postural challenge augmented the cerebrovascular coupling strength. This result might be the consequence of the raise of AP variability [64,65] that is not buffered with suitable changes to the vessel diameter and drives changes of MCBv via the pressure-to-flow relationship [66,67]. This consideration suggests a tendency toward a worsening of the dCA given that an increased association between MAP and MCBv variations has been observed during fainting [68] and in subjects with impaired dCA [9,69–71]. A linear approach based on $K^2_{MAP,MCBv}(f)$ suggested an increased cerebrovascular coupling strength as well. However, it is worth noting that the observed increase is likely to occur in a physiological range that does not compromise the dCA given that the quality of the MCBv response to a sustained step increase of MAP was found to be preserved during HUT [72,73]. Regardless of the magnitude of the changes, the robust assessment of the MCBv–MAP coupling strength is a critical issue. Indeed, traditional markers of MCBv–MAP association based on $K^2_{MAP,MCBv}(f)$ failed to detect an increased cerebrovascular coupling strength in pathological subjects with impaired dCA [74].

Remarkably, CSampEn did not decrease during HUT, thus stressing its limited ability to quantify the modifications of the cerebrovascular coupling strength during orthostatic challenge. This result might be the consequence of the causal structure of the dynamic MCBv–MAP interactions occurring in a closed loop [45,54,55,75] according to the pressure-to-flow link [76] and the Cushing-like pathway [77] mediated by the activity of the mechanoreceptors of the brainstem [78] and fully operative under physiological modifications of intracranial pressure [79].

7. Conclusions

This study proves the different abilities of CSampEn and KNNCUP metrics in assessing the coupling strength between two time series and clarifies the conditions under which their behaviors diverge. In addition to the trivial disruption of the state space correspondence resulting from nonlinear interactions occurring according to the $n{:}m$ coupling ratio, with m and n different from 1, even interactions occurring according to a causal structure are sufficient to limit the performance of CSampEn. We recommend the avoidance of the use of CSampEn to assess cardiorespiratory and cerebrovascular coupling from spontaneous variability, because CSampEn might be inadequate to describe the complex causal nature of these relationships [4–6,42,44,45,55,62,75,80]. Conversely, KNNCUP assures a more reliable nonlinear model-free framework to quantify cardiorespiratory and cerebrovascular variability interactions because its performances are less dependent on the structure of the underlying mechanism generating the dynamics, thus allowing the detection of a trend toward an increase of the cardiorespiratory coupling strength while slowing the breathing rate and the raise of the cerebrovascular coupling strength during the sympathetic activation and vagal withdrawal induced by an orthostatic challenge.

Author Contributions: Conceptualization, A.P.; methodology, A.P.; software, A.P.; validation, A.P.; formal analysis, A.P., V.B., F.G., B.C. and B.D.M.; investigation, D.T. and G.R.; resources, A.P., V.B. and B.C.; data curation, V.B., F.G. and B.C.; writing—original draft preparation, A.P.; writing—review and editing, A.P., V.B., F.G., B.C., B.D.M., D.T., G.R. and L.F.; visualization, A.P.; supervision, A.P.; project administration, A.P., V.B. and B.C.; funding acquisition, A.P., V.B. and B.C. All authors have read and agreed to the published version of the manuscript.

Funding: This research was funded by the Italian Ministry of Health grant RF-2016-02361069 (project PEARL, principal investigator: A.P.) to Policlinico San Donato.

Institutional Review Board Statement: The study was conducted in accordance with the Declaration of Helsinki, and approved by the Ethics Committees of L. Sacco Hospital, Milan, Italy, (CB protocol code: 1999-3; date of approval: 1/2/1999) and Sacro Cuore Don Calabria Hospital, Negrar, Italy (HUT protocol code: 101/2010; date of approval: 1/5/2010).

Data Availability Statement: The data presented in this study are available on request from the corresponding author (i.e., A.P.) upon permission of the IRCCS Policlinico San Donato. The data are not publicly available because they contain information that could compromise the privacy of research participants.

Conflicts of Interest: The authors declare no conflict of interest. The funders had no role in the design of the study; in the collection, analyses, or interpretation of data; in the writing of the manuscript, or in the decision to publish the results.

References

1. Elstad, M.; O'Callaghan, E.L.; Smith, A.J.; Ben-Tal, A.; Ramchandra, R. Cardiorespiratory interactions in humans and animals: Rhythms for life. *Am. J. Physiol.* **2018**, *315*, H6–H17. [CrossRef] [PubMed]
2. Penzel, T.; Kantelhardt, J.W.; Bartsch, R.P.; Riedl, M.; Kramer, J.; Wessel, N.; Garcia, C.; Glos, M.; Fietze, I.; Schöbel, C. Modulations of heart rate, ECG, and cardio-respiratory coupling observed in polysomnography. *Front. Physiol.* **2016**, *7*, 460. [CrossRef] [PubMed]
3. Yana, K.; Saul, J.P.; Berger, R.D.; Perrott, M.H.; Cohen, R.J. A time domain approach for the fluctuation analysis of heart rate related to instantaneous lung volume. *IEEE Trans. Biomed. Eng.* **1993**, *40*, 74–81. [CrossRef] [PubMed]
4. Saul, J.P.; Berger, R.D.; Chen, M.H.; Cohen, R.J. Transfer function analysis of autonomic regulation II. Respiratory sinus arrhythmia. *Am. J. Physiol.* **1989**, *256*, H153–H161. [CrossRef]
5. Triedman, J.K.; Perrott, M.H.; Cohen, R.J.; Saul, J.P. Respiratory sinus arrhythmia: Time domain characterization using autoregressive moving average analysis. *Am. J. Physiol.* **1995**, *268*, H2232–H2238. [CrossRef]
6. Porta, A.; Bassani, T.; Bari, V.; Tobaldini, E.; Takahashi, A.C.M.; Catai, A.M.; Montano, N. Model-based assessment of baroreflex and cardiopulmonary couplings during graded head-up tilt. *Comput. Biol. Med.* **2012**, *42*, 298–305. [CrossRef]
7. Porta, A.; Maestri, R.; Bari, V.; De Maria, B.; Cairo, B.; Vaini, E.; La Rovere, M.T.; Pinna, G.D. Paced breathing increases the redundancy of cardiorespiratory control in healthy individuals and chronic heart failure patients. *Entropy* **2018**, *20*, 949. [CrossRef]

8. Claassen, J.A.; Meel-van den Abeelen, A.S.; Simpson, D.M.; Panerai, R.B. and the international Cerebral Autoregulation Research Network (CARNet). Transfer function analysis of dynamic cerebral autoregulation: A white paper from the International Cerebral Autoregulation Research Network. *J. Cereb. Blood Flow Metab.* **2016**, *36*, 665–680. [CrossRef]
9. Giller, C.A. The frequency-dependent behavior of cerebral autoregulation. *Neurosurgery* **1990**, *27*, 362–368. [CrossRef]
10. Zhang, R.; Zuckerman, J.H.; Iwasaki, K.; Wilson, T.E.; Crandall, C.G.; Levine, B.D. Autonomic neural control of dynamic cerebral autoregulation in humans. *Circulation* **2002**, *106*, 1814–1820. [CrossRef]
11. Tzeng, Y.C.; Ainslie, P.N.; Cooke, W.H.; Peebles, K.C.; Willie, C.K.; Macrae, B.A.; Smirl, J.D.; Horsman, H.M.; Rickards, C.A. Assessment of cerebral autoregulation: The quandary of quantification. *Am. J. Physiol.* **2012**, *303*, H658–H671. [CrossRef]
12. Vaini, E.; Bari, V.; Fantinato, A.; Pistuddi, V.; Cairo, B.; De Maria, B.; Ranucci, M.; Porta, A. Causality analysis reveals the link between cerebrovascular control and acute kidney dysfunction after coronary artery bypass grafting. *Physiol. Meas.* **2019**, *40*, 064006. [CrossRef] [PubMed]
13. Hori, D.; Nomura, Y.; Ono, M.; Joshi, B.; Mandal, K.; Cameron, D.; Kocherginsky, M.; Hogue, C.W. Optimal blood pressure during cardiopulmonary bypass defined by cerebral autoregulation monitoring. *J. Thorac. Cardiovasc. Surg.* **2017**, *154*, 1590–1598. [CrossRef] [PubMed]
14. Porta, A.; Gelpi, F.; Bari, V.; Cairo, B.; De Maria, B.; Panzetti, C.M.; Cornara, N.; Bertoldo, E.G.; Fiolo, V.; Callus, E.; et al. Monitoring the evolution of asynchrony between mean arterial pressure and mean cerebral blood flow via cross-entropy methods. *Entropy* **2022**, *24*, 80. [CrossRef] [PubMed]
15. Eckberg, D.L.; Kifle, Y.T.; Roberts, V.L. Phase relationship between normal human respiration and baroreflex responsiveness. *J. Physiol.* **1980**, *304*, 489–502. [CrossRef]
16. Taha, B.H.; Simon, P.M.; Dempsey, J.A.; Skatrud, J.B.; Iber, C. Respiratory sinus arrhythmia in humans: An obligatory role for vagal feedback from the lungs. *J. Appl. Physiol.* **1995**, *78*, 638–645. [CrossRef]
17. Crystal, G.J.; Salem, M.R. The Bainbridge and the "reverse" Bainbridge reflexes: History, physiology, and clinical relevance. *Anesth. Analg.* **2012**, *114*, 520–532. [CrossRef]
18. Levy, M.N. Sympathetic-parasympathetic interactions in the heart. *Circ. Res.* **1971**, *29*, 437–445. [CrossRef]
19. Kawada, T.; Sugimachi, M.; Shishido, T.; Miyano, H.; Sato, T.; Yoshimura, R.; Miyashita, H.; Nakahara, T.; Alexander, J.; Sunagawa, K. Simultaneous identification of static and dynamic vagosympathetic interactions in regulating heart rate. *Am. J. Physiol.* **1999**, *276*, R782–R789. [CrossRef]
20. Taylor, J.A.; Myers, C.W.; Halliwill, J.R.; Seidel, H.; Eckberg, D.L. Sympathetic restraint of respiratory sinus arrhythmia: Implications for vagal-cardiac tone assessment in humans. *Am. J. Physiol.* **2001**, *280*, H2804–H2814. [CrossRef]
21. Angelone, A.; Coulter, N.A. Respiratory sinus arrhythmia: A frequency dependent phenomenon. *J. Appl. Physiol.* **1964**, *19*, 479–482. [CrossRef] [PubMed]
22. Hirsch, J.A.; Bishop, B. Respiratory sinus arrhythmia in humans: How breathing pattern modulates heart rate. *Am. J. Physiol.* **1981**, *241*, H620–H629. [CrossRef] [PubMed]
23. Eckberg, D.L. Human sinus arrhythmia as an index of vagal cardiac outflow. *J. Appl. Physiol.* **1983**, *54*, 961–966. [CrossRef] [PubMed]
24. Schäfer, G.; Rosenblum, M.G.; Kurths, J.; Abel, H.H. Heartbeat synchronized with ventilation. *Nature* **1998**, *392*, 239–240. [CrossRef]
25. Cairo, B.; Martins de Abreu, R.; Bari, V.; Gelpi, F.; De Maria, B.; Rehder-Santos, P.; Sakaguchi, C.A.; Donisete da Silva, C.; De Favari Signini, E.; Catai, A.M.; et al. Optimizing phase variability threshold for automated synchrogram analysis of cardiorespiratory interactions in amateur cyclists. *Philos. Trans. R. Soc. A* **2021**, *379*, 20200251. [CrossRef]
26. Tzeng, Y.C.; Larsen, P.D.; Galletly, D.C. Cardioventilatory coupling in resting human subjects. *Exp. Physiol.* **2003**, *88*, 775–782. [CrossRef]
27. Spyer, K.M. Central nervous mechanisms responsible for cardio-respiratory homeostasis. *Adv. Exp. Med. Biol.* **1995**, *381*, 73–79.
28. Gilbey, M.P.; Jordan, D.; Richter, D.W.; Spyer, K.M. Synaptic mechanisms involved in the inspiratory modulation of vagal cardio-inhibitory neurones in the cat. *J. Physiol.* **1984**, *356*, 65–78. [CrossRef]
29. Seals, D.R.; Suwarno, N.O.; Dempsey, J.A. Influence of lung volume on sympathetic nerve discharge in normal subjects. *Circ. Res.* **1990**, *67*, 130–141. [CrossRef]
30. Eckberg, D.L.; Nerhed, C.; Wallin, B.G. Respiratory modulation of muscle sympathetic and vagal cardiac outflow in man. *J. Physiol.* **1985**, *365*, 181–196. [CrossRef]
31. Lassen, N.A. Cerebral blood flow and oxygen consumption in man. *Physiol. Rev.* **1959**, *39*, 183–238. [CrossRef] [PubMed]
32. Aaslid, R.; Blaha, M.; Sviri, G.; Douville, C.M.; Newell, D.W. Asymmetric dynamic cerebral autoregulatory response to cyclic stimuli. *Stroke* **2007**, *38*, 1465–1469. [CrossRef] [PubMed]
33. Schmidt, B.; Klingelhofer, J.; Perkes, I.; Czosnyka, M. Cerebral autoregulatory response depends on the direction of change in perfusion pressure. *J. Neurotrauma* **2009**, *26*, 651–656. [CrossRef]
34. Bari, V.; Marchi, A.; De Maria, B.; Rossato, G.; Nollo, G.; Faes, L.; Porta, A. Nonlinear effects of respiration on the crosstalk between cardiovascular and cerebrovascular control systems. *Philos. Trans. R. Soc. A* **2016**, *374*, 20150179. [CrossRef]
35. Panerai, R.B. The critical closing pressure of the cerebral circulation. *Med. Eng. Phys.* **2003**, *25*, 621–632. [CrossRef]
36. Hamner, J.W.; Tan, C.O.; Lee, K.; Cohen, M.A.; Taylor, J.A. Sympathetic control of the cerebral vasculature in humans. *Stroke* **2010**, *41*, 102–109. [CrossRef] [PubMed]

37. Gebber, G.L.; Barman, S.M. Brain stem vasomotor circuits involved in the genesis and entrainment of sympathetic nervous rhythms. *Progr. Brain Res.* **1977**, *47*, 61–75.
38. Marchi, A.; Bari, V.; De Maria, B.; Esler, M.; Lambert, E.; Baumert, M.; Porta, A. Simultaneous characterization of sympathetic and cardiac arms of the baroreflex through sequence techniques during incremental head-up tilt. *Front. Physiol.* **2016**, *7*, 438. [CrossRef]
39. Bartsch, R.P.; Kantelhardt, J.W.; Penzel, T.; Havlin, S. Experimental evidence for phase synchronization transitions in the human cardiorespiratory system. *Phys. Rev. Lett.* **2007**, *98*, 054102. [CrossRef]
40. Richman, J.S.; Moorman, J.R. Physiological time-series analysis using approximate entropy and sample entropy. *Am. J. Physiol.* **2000**, *278*, H2039–H2049. [CrossRef]
41. Porta, A.; Faes, L.; Bari, V.; Marchi, A.; Bassani, T.; Nollo, G.; Perseguini, N.M.; Milan, J.; Minatel, V.; Borghi-Silva, A.; et al. Effect of age on complexity and causality of the cardiovascular control: Comparison between model-based and model-free approaches. *PLoS ONE* **2014**, *9*, e89463. [CrossRef] [PubMed]
42. Porta, A.; Guzzetti, S.; Montano, N.; Pagani, M.; Somers, V.; Malliani, A.; Baselli, G.; Cerutti, S. Information domain analysis of cardiovascular variability signals: Evaluation of regularity, synchronisation and co-ordination. *Med. Biol. Eng. Comput.* **2000**, *38*, 180–188. [CrossRef] [PubMed]
43. Porta, A.; Baselli, G.; Guzzetti, S.; Pagani, M.; Malliani, A.; Cerutti, S. Prediction of short cardiovascular variability signals based on conditional distribution. *IEEE Trans. Biomed. Eng.* **2000**, *47*, 555–1564.
44. Faes, L.; Porta, A.; Rossato, G.; Adami, A.; Tonon, D.; Corica, A.; Nollo, G. Investigating the mechanisms of cardiovascular and cerebrovascular regulation in orthostatic syncope through an information decomposition strategy. *Auton. Neurosci. Basic Clin.* **2013**, *178*, 76–82. [CrossRef]
45. Bari, V.; De Maria, B.; Mazzucco, C.E.; Rossato, G.; Tonon, D.; Nollo, G.; Faes, L.; Porta, A. Cerebrovascular and cardiovascular variability interactions investigated through conditional joint transfer entropy in subjects prone to postural syncope. *Physiol. Meas.* **2017**, *38*, 976–991. [CrossRef] [PubMed]
46. Porta, A.; Bari, V.; Gelpi, F.; Cairo, B.; De Maria, B.; Tonon, D.; Rossato, G.; Faes, L. Comparing cross-sample entropy and k-nearest-neighbor cross-predictability approaches for the evaluation of cardiorespiratory and cerebrovascular dynamic interactions. In Proceedings of the 44th Annual International Conference of the IEEE EMBS, Glasgow, Scotland, 11–15 July 2022; IEEE Press: Piscataway, NJ, USA, 2022; pp. 127–130.
47. Schiff, S.J.; So, P.; Chang, T.; Burke, R.; Sauer, T. Detecting dynamical interdependence and generalized synchrony through mutual prediction in a neural ensemble. *Phys. Rev. E* **1996**, *54*, 6708–6724. [CrossRef]
48. Abarbanel, H.D.I.; Carroll, T.L.; Pecora, L.M.; Sidorowich, J.J.; Tsimring, L.S. Predicting physical variables in time-delay embedding. *Phys. Rev. E* **1994**, *49*, 1840–1853. [CrossRef]
49. Porta, A.; Castiglioni, P.; Bari, V.; Bassani, T.; Marchi, A.; Cividjian, A.; Quintin, L.; Di Rienzo, M. K-nearest-neighbor conditional entropy approach for the assessment of short-term complexity of cardiovascular control. *Physiol. Meas.* **2013**, *34*, 17–33. [CrossRef]
50. Lloyd, A.L. The coupled logistic map: A simple model for the effects of spatial heterogeneity on population dynamics. *J. Theor. Biol.* **1995**, *173*, 217–230. [CrossRef]
51. Aaslid, R.; Markwalder, T.M.; Nornes, H. Noninvasive transcranial Doppler ultrasound recording of flow velocity in basal cerebral arteries. *J. Neurosurg.* **1982**, *57*, 769–774. [CrossRef]
52. Task Force of the European Society of Cardiology and the North American Society of Pacing and Electrophysiology. Heart rate variability: Standards of measurement, physiological interpretation and clinical use. *Circulation* **1996**, *93*, 1043–1065. [CrossRef]
53. Porta, A.; Castiglioni, P.; Di Rienzo, M.; Bassani, T.; Bari, V.; Faes, L.; Nollo, G.; Cividjan, A.; Quintin, L. Cardiovascular control and time domain Granger causality: Insights from selective autonomic blockade. *Philos. Trans. R. Soc. A* **2013**, *371*, 20120161. [CrossRef] [PubMed]
54. Bari, V.; Fantinato, A.; Vaini, E.; Gelpi, F.; Cairo, B.; De Maria, B.; Pistuddi, V.; Ranucci, M.; Porta, A. Impact of propofol general anesthesia on cardiovascular and cerebrovascular closed loop variability interactions. *Biomed. Signal Process. Control* **2021**, *68*, 102735. [CrossRef]
55. Porta, A.; Gelpi, F.; Bari, V.; Cairo, B.; De Maria, B.; Tonon, D.; Rossato, G.; Ranucci, M.; Faes, L. Categorizing the role of respiration in cardiovascular and cerebrovascular variability interactions. *IEEE Trans. Biomed. Eng.* **2022**, *69*, 2065–2076. [CrossRef] [PubMed]
56. Porta, A.; Bari, V.; De Maria, B.; Cairo, B.; Vaini, E.; Perseguini, N.M.; Milan-Mattos, J.; Rehder-Santos, P.; Minatel, V.; Takahashi, A.C.M.; et al. Comparison between probabilistic and Wiener-Granger causality in assessing modifications of the cardiac baroreflex control with age. *Physiol. Meas.* **2018**, *39*, 104004. [CrossRef]
57. Faes, L.; Porta, A.; Nollo, G. Information decomposition in bivariate systems: Theory and application to cardiorespiratory dynamics. *Entropy* **2015**, *17*, 277–303. [CrossRef]
58. Porta, A.; Guzzetti, S.; Furlan, R.; Gnecchi-Ruscone, T.; Montano, N.; Malliani, A. Complexity and nonlinearity in short-term heart period variability: Comparison of methods based on local nonlinear prediction. *IEEE Trans. Biomed. Eng.* **2007**, *54*, 94–106. [CrossRef]
59. Chess, G.F.; Calaresu, F.R. Frequency response model of vagal control of heart rate in the cat. *Am. J. Physiol.* **1971**, *220*, 554–557. [CrossRef]
60. Berger, R.D.; Saul, J.P.; Cohen, R.J. Transfer function analysis of autonomic regulation I. Canine atrial rate response. *Am. J. Physiol.* **1989**, *256*, H142–H152. [CrossRef]

61. Chen, X.; Mukkamala, R. Selective quantification of the cardiac sympathetic and parasympathetic nervous systems by multisignal analysis of cardiorespiratory variability. *Am. J. Physiol.* **2008**, *294*, H362–H371. [CrossRef]
62. Baselli, G.; Cerutti, S.; Badilini, F.; Biancardi, L.; Porta, A.; Pagani, M.; Lombardi, F.; Rimoldi, O.; Furlan, R.; Malliani, A. Model for the assessment of heart period and arterial pressure variability interactions and respiratory influences. *Med. Biol. Eng. Comput.* **1994**, *32*, 143–152. [CrossRef] [PubMed]
63. Patton, D.J.; Triedman, J.K.; Perrott, M.H.; Vidian, A.A.; Saul, J.P. Baroreflex gain: Characterization using autoregressive moving average analysis. *Am. J. Physiol.* **1996**, *270*, H1240–H1249. [CrossRef] [PubMed]
64. Cooke, W.H.; Hoag, J.B.; Crossman, A.A.; Kuusela, T.A.; Tahvanainen, K.U.; Eckberg, D.L. Human responses to upright tilt: A window on central autonomic integration. *J. Physiol.* **1999**, *517*, 617–628. [CrossRef]
65. Marchi, A.; Bari, V.; De Maria, B.; Esler, M.; Lambert, E.; Baumert, M.; Porta, A. Calibrated variability of muscle sympathetic nerve activity during graded head-up tilt in humans and its link with noradrenaline data and cardiovascular rhythms. *Am. J. Physiol.* **2016**, *310*, R1134–R1143. [CrossRef]
66. Zhang, R.; Zuckerman, J.H.; Levine, B.D. Deterioration of cerebral autoregulation during orthostatic stress: Insights from the frequency domain. *J. Appl. Physiol.* **1998**, *85*, 1113–1122. [CrossRef] [PubMed]
67. Katsogridakis, E.; Bush, G.; Fan, L.; Birch, A.A.; Simpson, D.M.; Allen, R.; Potter, J.F.; Panerai, R.B. Detection of impaired cerebral autoregulation improves by increasing arterial blood pressure variability. *J. Cereb. Blood Flow Metab.* **2013**, *33*, 519–523. [CrossRef]
68. Ocon, A.J.; Kulesa, J.; Clarke, D.; Taneja, I.; Medow, M.S.; Stewart, J.M. Increased phase synchronization and decreased cerebral autoregulation during fainting in the young. *Am. J. Physiol.* **2009**, *297*, H2084–H2095. [CrossRef] [PubMed]
69. Ocon, A.J.; Medow, M.S.; Taneja, I.; Clarke, D.; Stewart, J.M. Decreased upright cerebral blood flow and cerebral autoregulation in normocapnic postural tachycardia syndrome. *Am. J. Physiol.* **2009**, *297*, H664–H673. [CrossRef]
70. Castro, P.; Serrador, J.; Sorond, F.; Azevedo, E.; Rocha, I. Sympathovagal imbalance in early ischemic stroke is linked to impaired cerebral autoregulation and increased infarct volumes. *Auton. Neurosci. Basic Clin.* **2022**, *241*, 102986. [CrossRef]
71. Czosnyka, M.; Smielewski, P.; Kirkpatrick, P.; Menon, D.K.; Pickard, J.D. Monitoring of cerebral autoregulation in head-injured patients. *Stroke* **1996**, *27*, 1829–1834. [CrossRef]
72. Carey, B.J.; Manktelow, B.N.; Panerai, R.B.; Potter, J.F. Cerebral autoregulatory responses to head-up tilt in normal subjects and patients with recurrent vasovagal syncope. *Circulation* **2001**, *104*, 898–902. [CrossRef] [PubMed]
73. Gelpi, F.; Bari, V.; Cairo, B.; De Maria, B.; Tonon, D.; Rossato, G.; Faes, L.; Porta, A. Dynamic cerebrovascular autoregulation in patients prone to postural syncope: Comparison of techniques assessing the autoregulation index from spontaneous variability series. *Auton. Neurosci. Basic Clin.* **2022**, *237*, 102920. [CrossRef] [PubMed]
74. Caldas, J.R.; Panerai, R.B.; Haunton, V.J.; Almeida, J.P.; Ferreira, G.S.R.; Camara, L.; Nogueira, R.C.; Bor-Seng-Shu, E.; Oliveira, M.L.; Groehs, R.R.V.; et al. Cerebral blood flow autoregulation in ischemic heart failure. *Am. J. Physiol.* **2017**, *312*, R108–R113. [CrossRef] [PubMed]
75. Saleem, S.; Teal, P.D.; Howe, C.A.; Tymko, M.M.; Ainslie, P.N.; Tzeng, Y.-C. Is the Cushing mechanism a dynamic blood pressure-stabilizing system? Insights from Granger causality analysis of spontaneous blood pressure and cerebral blood flow. *Am. J. Physiol.* **2018**, *315*, R484–R495. [CrossRef]
76. Tzeng, Y.C.; MacRae, B.A.; Ainslie, P.N.; Chan, G.S.H. Fundamental relationships between blood pressure and cerebral blood flow in humans. *J. Appl. Physiol.* **2014**, *117*, 1037–1048. [CrossRef] [PubMed]
77. Cushing, H. Concerning a definitive regulatory mechanism of the vaso-motor centre which controls blood pressure during cerebral compression. *Bull. Johns Hopkins Hosp.* **1901**, *12*, 290–292.
78. McBryde, F.D.; Malpas, S.C.; Paton, J.F.R. Intracranial mechanisms for preserving brain blood flow in health and disease. *Acta Physiol.* **2017**, *219*, 274–287. [CrossRef]
79. Nakamura, K.; Osborn, J.W.; Cowley, A.W. Pressor response to small elevations of cerebroventricular pressure in conscious rats. *Hypertension* **1987**, *10*, 635–641. [CrossRef]
80. Panerai, R.B.; Dawson, S.L.; Potter, J.F. Linear and nonlinear analysis of human dynamic cerebral autoregulation. *Am. J. Physiol.* **1999**, *277*, H1089–H1099. [CrossRef]

entropy

MDPI

Article

Induced Relaxation Enhances the Cardiorespiratory Dynamics in COVID-19 Survivors

Alejandra Margarita Sánchez-Solís [1], Viridiana Peláez-Hernández [2,*], Laura Mercedes Santiago-Fuentes [1,3,*], Guadalupe Lizzbett Luna-Rodríguez [2], José Javier Reyes-Lagos [1] and Arturo Orea-Tejeda [2]

1 School of Medicine, Universidad Autónoma del Estado de México (UAEMéx), Toluca de Lerdo 50180, Mexico; asanchezs199@alumno.uaemex.mx (A.M.S.-S.); jjreyesl@uaemex.mx (J.J.R.-L.)

2 Cardiology Service, Instituto Nacional de Enfermedades Respiratorias Ismael Cosío Villegas (INER), Mexico City 14080, Mexico; g.lizzbett.luna.rod@gmail.com (G.L.L.-R.); oreatart@gmail.com (A.O.-T.)

3 Health Sciences Department, Universidad Autónoma Metropolitana Unidad Iztapalapa (UAM-I), Mexico City 09340, Mexico

* Correspondence: vpelaezh@gmail.com (V.P.-H.); lmsf@xanum.uam.mx (L.M.S.-F.)

Abstract: Most COVID-19 survivors report experiencing at least one persistent symptom after recovery, including sympathovagal imbalance. Relaxation techniques based on slow-paced breathing have proven to be beneficial for cardiovascular and respiratory dynamics in healthy subjects and patients with various diseases. Therefore, the present study aimed to explore the cardiorespiratory dynamics by linear and nonlinear analysis of photoplethysmographic and respiratory time series on COVID-19 survivors under a psychophysiological assessment that includes slow-paced breathing. We analyzed photoplethysmographic and respiratory signals of 49 COVID-19 survivors to assess breathing rate variability (BRV), pulse rate variability (PRV), and pulse–respiration quotient (PRQ) during a psychophysiological assessment. Additionally, a comorbidity-based analysis was conducted to evaluate group changes. Our results indicate that all BRV indices significantly differed when performing slow-paced breathing. Nonlinear parameters of PRV were more appropriate for identifying changes in breathing patterns than linear indices. Furthermore, the mean and standard deviation of PRQ exhibited a significant increase while sample and fuzzy entropies decreased during diaphragmatic breathing. Thus, our findings suggest that slow-paced breathing may improve the cardiorespiratory dynamics of COVID-19 survivors in the short term by enhancing cardiorespiratory coupling via increased vagal activity.

Keywords: cardiorespiratory coupling; post-COVID-19 syndrome; slow breathing; diaphragmatic breathing; breathing and relaxation exercises; pulse–respiration quotient

1. Introduction

The COVID-19 disease, caused by the novel coronavirus SARS-CoV-2, emerged at the end of 2019 and has rapidly spread worldwide since it first appeared in China, resulting in more than 673 million cases and 6.8 million deaths at the time of writing this article [1–3]. The pathological response to the infection ranges from asymptomatic to severe cases, and the main complications caused by the disease include hypoxemic respiratory failure, heart failure, kidney injury, thromboembolic damage, and acute respiratory distress syndrome (ARDS), which represents most ICU admissions due to COVID-19 illness [4–6]. A significant proportion of individuals reported persistent symptoms beyond 12 weeks and up to 9 months that were not associated with any other diagnosis and occurred after the recovery from the acute phase of the disease [7,8], regardless of if the patient experienced an asymptomatic or severe case [9]. These manifestations are part of the post-acute sequelae of SARS-CoV-2 infection (PASC) or post-COVID syndrome, which involves respiratory, neurological, endocrine, cardiovascular, and other multisystemic alterations [10,11].

A high prevalence of autonomic alterations in respiratory and cardiovascular systems, which modify cardiopulmonary interactions as a sign of dysautonomia or orthostatic syndromes, occurs in subjects who suffered an acute phase of the disease [12–14]. Several signs and symptoms associated with this autonomic impairment have been reported, with chronic fatigue, chest pain, dyspnea, sleep disturbances, anxiety, depression, and post-traumatic stress syndrome being the most common among post-COVID patients [15–18]. Recent research on the dynamics of the autonomic nervous system (ANS) in pathological conditions has demonstrated the existence of an autonomic imbalance in orthostatic syndrome [19], respiratory disease [20], and cardiovascular pathologies [21] by mathematical modeling of physiological interactions related to system coupling and causality. However, scarce studies have addressed the study of ANS dynamics in COVID survivors. In this regard, the pulse–respiration quotient (PRQ) has proven to be a powerful parametrization tool for analyzing the complex behavior of the ANS and its interactions in a wide range of clinical and research settings [22].

Slow-paced breathing has been introduced as a relaxation technique that consists of reducing the breathing rate (BR) to about 6 cycles per minute (0.1 Hz), usually with 5-s deep inhalation and exhalation phases [23]. This breathing technique, also referred to as diaphragmatic [24] or abdominal slow-paced breathing [25], has shown a beneficial impact in both healthy and ill individuals [26,27]. Other breathing exercises, such as meditation and mindfulness, have been shown to reduce depression, pain, anxiety, and stress indicators [28]. Recent reports have suggested that increased respiratory sinus arrhythmia (RSA) mediated by a lower breathing rate improves baroreflex sensitivity and increases parasympathetic nervous control of the heart [29,30]. However, there is limited evidence on the effects of psychophysiological assessments on the ANS response of COVID-19 survivors, and the ongoing studies do not include individuals with concomitant disease [31]. A relevant study highlights the importance of assessing autonomic function to detect possible alterations that may compromise the recovery of COVID-19 survivors and to design biofeedback training based on their characteristics [32].

Thus, this work aimed to study the cardiorespiratory dynamics by linear and nonlinear analysis of photoplethysmographic and respiratory time series on COVID-19 survivors under a psychophysiological assessment (study I). In addition, as a secondary exploratory study, the effect of different comorbidities on the cardiorespiratory dynamics of COVID-19 survivors was examined (study II). We hypothesized that relaxation induces increased cardiorespiratory parasympathetic activity in COVID-19 survivors.

2. Materials and Methods

2.1. Subjects

This exploratory study involved 49 survivors of COVID-19 illness (32 men and 17 women) aged 27 to 84 years old (49 ± 12 years) who were hospitalized between 2020 and 2022 at the Instituto Nacional de Enfermedades Respiratorias (INER, Mexico City) due to COVID-19 complications in the acute infection phase. All the participants were laboratory-confirmed with SARS-CoV-2 and surveyed at three months post-hospital discharge. Table 1 depicts the demographic and clinical characteristics of the sample; psychological disorders such as stress, anxiety, and depression were the most common among participants (men: 9; women: 11) who disclosed having sequelae after the infection, while most patients reported at least one cardiovascular, pulmonary, motor, cognitive, or other symptoms in the aftermath of the disease.

The research protocol was approved by the Ethics Committee of the National Institute of Respiratory Diseases, Mexico (approval number C57-21). All volunteers in this study signed an Informed Consent form when they agreed to participate, and all methods were performed following the relevant guidelines and regulations. The exclusion criteria involved limited mobility and receiving treatment at home instead of being hospitalized. The sample size for this exploratory research was determined by convenience sampling.

Table 1. Demographic and clinical characteristics of the patients.

Variable	Sample (n = 49)	Men (n = 32)	Women (n = 17)
No comorbidities	26 (53%)	15 (47%)	11 (65%)
≥1 comorbidity	23 (47%)	17 (53%)	6 (35%)
Mean hospital stay	21 days	22.4 days	19.5 days
No comorbidities	15.6 days	14.8 days	15.6 days
≥1 comorbidity	28 days	29 days	24.8 days
Sequelae	26 (53%)	14 (44%)	12 (71%)
Cardiovascular	2	1	1
Pulmonary	2	1	1
Motor	4	2	2
Psychological	20	11	9
Cognitive	4	2	2
Others	7	2	5

2.2. Psychophysiological Assessment (Protocol for Relaxation)

A validated protocol of 10 min for relaxation [32,33] was conducted on COVID-19 survivors in a quiet room in the Coordination of Psychology of the Cardiology Department of the INER between 8 a.m. and 12 p.m. to avoid major fluctuations associated with circadian rhythms [34,35]. The protocol for relaxation involved four phases with the same temporal duration: (1) open eyes; (2) closed eyes; (3) natural relaxation, and (4) induced relaxation. The specific duration and tasks of the participants in each phase are shown in Table 2. In addition, participants were asked to sit comfortably positioned sitting upright in a chair, resting their hands on their legs without taking their feet off the ground or moving while recording was performed.

Table 2. Protocol for relaxation applied to COVID-19 survivors.

	Phase	Duration	Tasks
1	**Open eyes**	2.5 min	Sitting upright with open eyes, spontaneous breathing
2	**Closed eyes**	2.5 min	Sitting upright with closed eyes, spontaneous breathing
3	**Natural relaxation**	2.5 min	Sitting upright with closed eyes, natural skill of relaxation
4	**Induced relaxation**	2.5 min	Sitting upright with closed eyes, slow-paced breathing (six breaths/min, 1:1 inhalation to exhalation ratio)

The COVID-19 survivors were instructed to follow all the phases in this protocol. Before the test, the medical staff guided the patient on how to perform a paced-breathing technique at a fixed rate of six breaths/minute with a 1:1 ratio of inhalation to exhalation time (5 s each). Additionally, for the natural relaxation phase, individuals were asked to relax without an explicit indication of modifying their breathing pattern or frequency as requested for diaphragmatic breathing. In this study, the protocol for relaxation belongs to a preliminary psychophysiological phase of examination of participants who subsequently were enrolled in biofeedback therapy based on their heart rate and breathing rate values.

2.3. Data Acquisition and Pre-Processing Process of Photoplethysmographic and Respiratory Time Series in COVID-19 Survivors

Simultaneously to the protocol for relaxation, the COVID-19 survivors were instrumented with a finger-clip pulse oximeter sensor on the thumb of the left hand, followed by two adjustable sensor belts, one on the chest and another on the abdomen (Figure 1). The sensors belong to a ProComp Infiniti System (Thought Technology Ltd., Montreal, CA, Canada) with BVP-Flex/Pro (photoplethysmography sensor/pulse oximeter) and

Resp-Flex/Pro (respiration sensor belt for chest and abdominal recordings). The pulse and respiratory signals were recorded with a sampling frequency of 2048 Hz and 256 Hz, respectively. The photoplethysmographic (PPG) and respiratory (RESP) signals were recorded without pauses among phases and had a total duration of 10 min, with 2.5 min for each task (Figures 1 and 2).

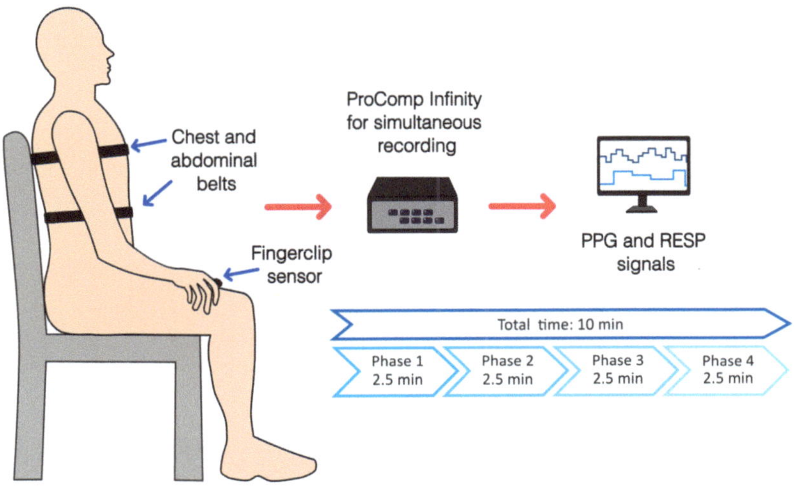

Figure 1. Placement of sensors for physiological recordings of photoplethysmography (PPG, finger clip pulse oximeter) and respiratory signals (RESP, chest, and abdominal belts) during the complete protocol for relaxation. An overview of the phases of the protocol and its duration is shown.

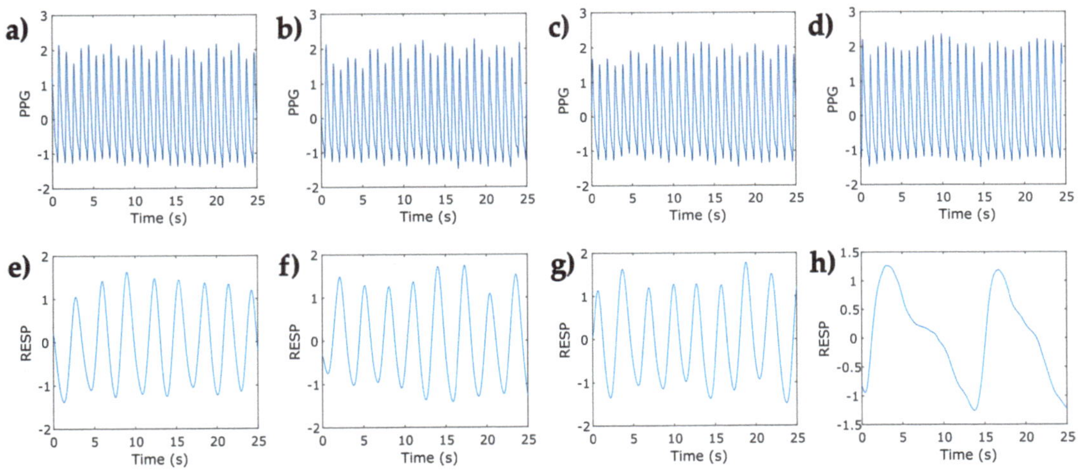

Figure 2. Representative segments (25 s) of photoplethysmographic (PPG) and respiratory signals (RESP) corresponding to one participant in the protocol for relaxation applied to COVID-19 survivors. (**a**) open eyes Phase—PPG signal; (**b**) closed eyes—PPG signal; (**c**) natural relaxation—PPG signal; (**d**) induced relaxation—PPG signal; (**e**) open eyes—RESP signal; (**f**) closed eyes—RESP signal; (**g**) natural relaxation—RESP signal; and (**h**) induced relaxation—RESP signal.

The physiological data recordings were exported as text files to Matlab R2022a (The MathWorks Inc., Natick, MA, USA). The extracted raw RESP signals were filtered with a second-order IIR Butterworth bandpass filter with a bandpass frequency range of 0.05 Hz to

1 Hz. Subsequently, a fourth-order IIR Butterworth low-pass filter with a cutoff frequency of 5 Hz was used to filter the raw PPG signals to remove noise interference.

2.4. Data Analysis

After the pre-processing stage, the PPG and RESP time series with more than 10% loss of information at any phase were discarded. In addition, RESP signals with an abnormal breathing pattern were not considered for further analyses. In consequence, forty-nine signals were included for breathing rate variability (BRV), pulse rate variability (PRV), and pulse–respiration quotient (PRQ) data analyses, as indicated in Figure 3.

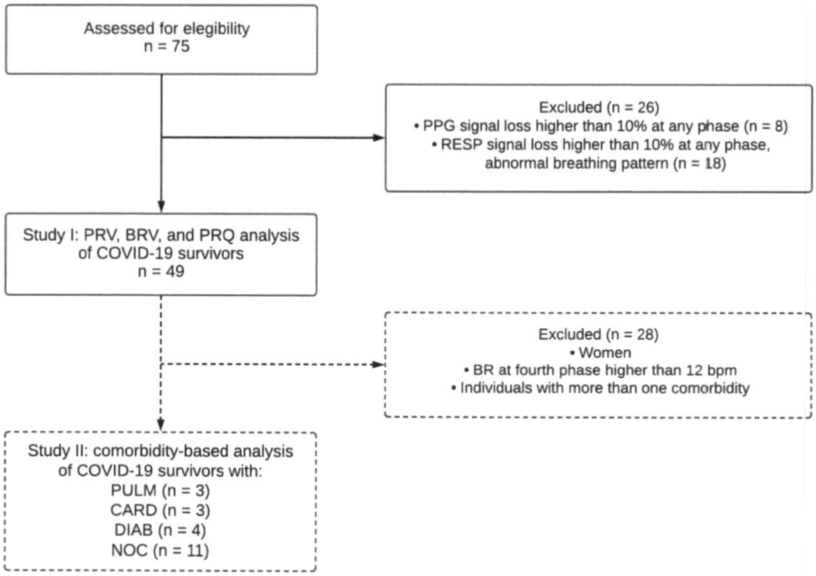

Figure 3. The flowchart describes the procedure for sample selection criteria for the photoplethysmographic (PPG) and respiratory (RESP) signals corresponding to COVID-19 survivors. A total of forty-nine records were included (study I) in the analysis of pulse rate variability (PRV), breathing rate variability (BRV), and pulse–respiration quotient (PRQ). Subsequently, a subsample of twenty-one individuals was evaluated for a comorbidity-based analysis (study II). Note: BR breathing rate; bpm beats per minute; PULM pulmonary disease; CARD cardiovascular disease; DIAB diabetes; NOC no comorbidities.

In this study, two main stages were assessed. In study I, forty-nine signals were included for breathing rate variability (BRV), pulse rate variability (PRV), and pulse–respiration quotient (PRQ) data analyses, as indicated in Figure 3. Furthermore, in study II, as a complementary and exploratory study, we separated the individuals by pre-existing comorbidities. To avoid the influence of sex hormones in the autonomic vagal activity predominantly found in young women [22,36], the group was reduced to 21 male individuals sorted by comorbidities into four categories: 1. patients with pulmonary disease (PULM), 2. patients with cardiovascular disease (CARD), 3. patients with diabetes (DIAB), and 4. individuals with no comorbidities (NOC).

2.4.1. Breathing Rate Variability (BRV)

Peak detection was performed to RESP signals from the local inspiration maxima of each phase to extract the breath-to-breath (BB) time series (expressed in seconds) [37]. The mean breathing rate (mean BR), the standard deviation of BB intervals (SDBB), and the root mean square of successive differences of BB intervals (RMSSD) were computed as proposed by Soni and Muniyandi [38].

2.4.2. Pulse Rate Variability (PRV)

A peak detection analysis was performed on PPG signals to obtain the pulse-to-pulse time series (IBI) expressed in milliseconds (ms); it is reported as pulse rate variability (PRV) since the data were obtained from pulse recordings instead of ECG signals [39,40]. The linear and nonlinear indices (see Table 3) were computed using the Kubios HRV Standard 3.5 software (Kubios Oy, Kuopio, Finland) [41].

Table 3. Parameters calculated from beat-to-beat interval (IBI) times series of COVID-19 survivors performing a protocol for relaxation.

Type	Index
Linear	Mean PP distance
	SDNN
	RMSSD
	LF power
	HF power
	LF/HF ratio
Nonlinear	SampEn
	DFA α_1
	DFA α_2

PP—pulse-to-pulse; SDNN—standard deviation of normal pulse intervals; RMSSD—root mean square of successive PP interval differences; LF—relative power of the low-frequency band; HF—relative power of the high-frequency band; LF/HF—ratio of low to high-frequency band power; SampEn—sample entropy; DFA α_1—detrended fluctuation analysis (short-term); DFA α_2—detrended fluctuation analysis (long-term). Adapted from [42].

2.4.3. Pulse–Respiration Quotient (PRQ)

The pulse–respiration quotient (PRQ) is a parameter reintroduced by Scholkmann and Wolf, obtained from the ratio of the heart rate to the breathing rate that effectively reflects the changes in the cardiorespiratory dynamics mediated by age, sex, physical and cognitive activity, body posture, and chronobiological state, and it is sensitive to the activity executed at the time of the recording [22]. It can be computed from the instantaneous heart rate and breathing rate, or the mean value of both parameters as follows:

$$\text{PRQ} = \frac{HR}{BR}\,\frac{beats/min}{breaths/min} \tag{1}$$

As proposed by Scholkmann and Wolf, all data regarding the PRQ index are reported as daytime functional PRQ (D-f-PRQ) [22] since our recordings were collected in the morning.

Instantaneous heart rate and breathing rate time series were obtained from IBI and BB time series, respectively. To ensure equidistant signals, the instantaneous breathing rate time series (Figure 4b) was upsampled to match the equivalent value of the breathing rate at the same instant of the beat occurrence (Figure 4a), both signals were then used for calculating the instantaneous PRQ. Figure 5 depicts the PRQ time series computation from the instantaneous heart rate (HR) and breathing rate (BR) signals for the four phases of the protocol for relaxation (Table 2).

The PRQ analysis included the mean PRQ and standard deviation of instantaneous PRQ (SD-PRQ) as the linear indexes, followed by sample entropy (SampEn) and fuzzy entropy (FuzzyEn) as the nonlinear parameters; the computing of the nonlinear indexes is described below.

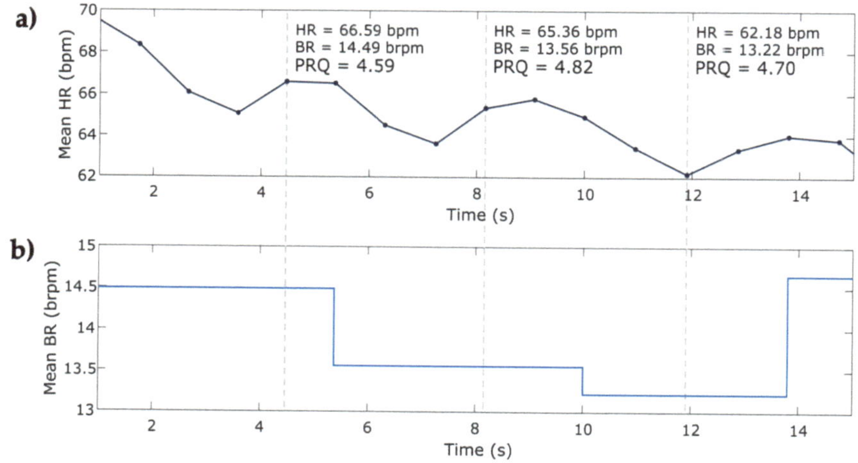

Figure 4. Visual representation of the computing of instantaneous PRQ index from the instantaneous heart and breathing rate from one subject at phase 1 of the protocol. (**a**) 15 s of the instantaneous heart rate time series; (**b**) 15 s of the upsampled instantaneous breathing rate time series. The figure shows HR heart rate; bpm beats per minute; BR breathing rate; brpm breaths per minute; pulse–respiration quotient (PRQ).

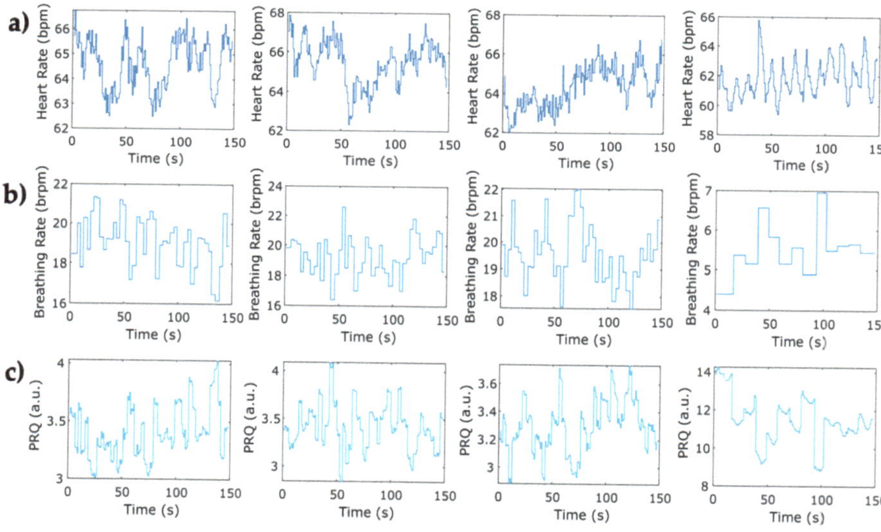

Figure 5. Representative examples of pulse–respiration quotient time series computed from instantaneous heart rate and breathing rate for the four phases of the protocol for relaxation in one COVID-19 survivor. (**a**) Instantaneous heart rate at the four phases obtained from a sample of photoplethysmographic record; (**b**) sample of the instantaneous breathing rate at four phases of the protocol obtained from breathing signals; (**c**) instantaneous pulse–respiration quotient computed from instantaneous HR and BR series (**a**,**b**).

Sample Entropy (SampEn)

Sample entropy is a widely known parameter used to evaluate how regular the fluctuations of a time series are, reducing the measured bias usually found when using approximate entropy (ApEn) associated with the signal length and self-matches [43]. This method described by Richman and Moorman is computed as follows:

From a vector $x_N = \{x_1, x_2, x_3, \ldots, x_N : 1 \leq i \leq N\}$ of N samples, the vectors $u(i)$ are formed as a series of m successive points:

$$u(i) = [x(i), x(i+1), x(i+2), \ldots, x(i+m-1)], \quad (2)$$

m being the embedding dimension of the vector length. From that, the distance d, defined as the maximum absolute difference of the elements of the vectors is computed; two points are similar when d between them is less than r:

$$d[u(i), u(j)] = \max\{|u(i+k) - u(j+k)| : 0 \leq k \leq m-1\} \leq r \quad (3)$$

where r is the tolerance factor and equals to 0.1–0.2 times the standard deviation SD of x_N when the data are not normalized [43–46].

Then, we calculate the number of times the distance of $u(j)$ from $u(i)$ vector is less or equal to r, for which $j \neq i$, $j \leq N-m+1$, to avoid self-matches:

$$C_i^m(r) = (N-m+1)^{-1} \sum_{i=1}^{N-m} number\ of\ times\ that\ d[u(i),\ u(j)] \leq r \quad (4)$$

The average similarity Φ is then determined as:

$$\Phi^m(r) = \frac{1}{N-m+1} \sum_{i=1}^{N-m+1} C_i^m(r). \quad (5)$$

Therefore, SampEn is obtained by:

$$SampEn(m, r, N) = -\ln\left(\frac{\Phi^m(r)}{\Phi^{m+1}(r)}\right) \quad (6)$$

Fuzzy Entropy (FuzzyEn)

Fuzzy entropy, introduced by Cheng et al. and based on the fuzzy set theory proposed by De Luca and Termini, evaluates the complexity and ambiguity of a given data set [47,48]. This measure is more reliable for short-length data and is less affected by noise [49]. FuzzyEn is computed as follows:

From a time series of N samples $\{u(i) : 1 \leq i \leq N\}$, vectors of m length are formed as follows:

$$X_i^m = \{u(i), u(i+1), \ldots,\ u(i+m-1)\} - u_0(i)\ ,\ i = 1\ , \ldots\ , N-m+1, \quad (7)$$

where X_i^m is conformed of m successive u values, and $u_0(i)$ represents the subtraction of the baseline of each vector to normalize the data:

$$u_0(i) = m^{-1} \sum_{j=0}^{m-1} u(i+j). \quad (8)$$

Afterward, d_{ij}^m is defined as the maximum absolute difference between the elements of X_i^m and X_j^m:

$$d_{ij}^m = d\left[X_i^m, X_j^m\right] = \max|(u(i+k) - u_0(i)) - (u(j+k) - u_0(j))|\ :\ 0 \leq k \leq m, \quad (9)$$

from which the similarity degree is obtained by the fuzzy function with given n and r parameters:

$$D_{ij}^{m}(n,r) = \mu\left(d_{ij}^{m}, n, r\right) = \exp\left(-\frac{\left(d_{1j}^{m}\right)^{n}}{r}\right), i \neq j \tag{10}$$

Then, the average similarity expressed by $\varphi^{m}(r)$ is computed:

$$\varphi^{m}(r) = \frac{1}{N-m}\sum_{i=1}^{N-m}\left(\frac{1}{N-m-1}\sum_{j=1,\, j\neq i}^{N-m} D_{ij}^{m}\right), \tag{11}$$

to later determine the FuzzyEn as:

$$FuzzyEn(m,n,r,N) = \ln \varphi^{m}(r) - \ln \varphi^{m+1}(r) \tag{12}$$

These entropies were evaluated using PyBios 1.0.0 software (Ribeirão Preto, SP, Brazil), setting the following parameters for sample entropy (m = 2, r = 0.2) and fuzzy entropy (m = 2, r = 0.2, n = 2) [50].

2.5. Statistical Analysis

The mean values of linear and nonlinear indices obtained from BB, IBI, and PRQ time series were evaluated by a Lilliefors (Kolmogorov–Smirnov) test to assess normal distribution. Subsequently, multiple comparisons between phases with respect to Phase 4 were performed; paired t-tests were conducted if data met the normality assumption; otherwise, we conducted a non-parametric Wilcoxon signed-rank test instead. A non-parametric Kruskal Wallis test was carried out for all indices extracted from BRV, PRV, and PRQ analysis at the induced relaxation phase for comorbidity-based statistical analysis. All these statistical tests were performed using Matlab R2022a software (The MathWorks Inc., Natick, MA, USA).

3. Results

3.1. Breathing Rate Variability

A lower significant mean value of the respiration rate was found in Phase 4 (9.3 ± 3.6 breaths/min) compared to the other three Phases (p < 0.001) following a slow breathing pattern (Figure 6a). It confirmed that individuals reduced the number of breaths per minute. Meanwhile, the SDBB and RMSSD increased significantly (p < 0.001) in Phase 4 compared to the other three Phases (Figure 6b,c).

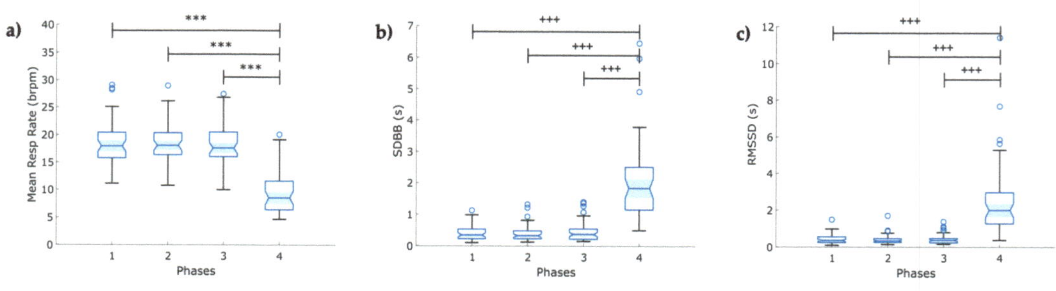

Figure 6. Boxplot of indices of the breathing rate variability (BRV) extracted from the respiration signal (RESP) in COVID-19 survivors (N = 49) under a psychophysiological assessment of four phases: (**a**) mean respiration rate; (**b**) standard deviation of breath-to-breath (BB) intervals (SDBB); and (**c**) root mean square of the successive difference between breath intervals (RMSSD). The figure shows: bpm breaths per minute, ∘ outliers, *** p < 0.001 obtained by paired t-test, +++ p < 0.001 obtained by Wilcoxon signed-rank test.

3.2. Pulse Rate Variability

3.2.1. Linear Features

A significant increase ($p < 0.001$) is exhibited for SDNN and LF/HF ratio in Phase 4 compared with the other three phases. Figure 7 compares the linear indexes of each phase from the PPG time series.

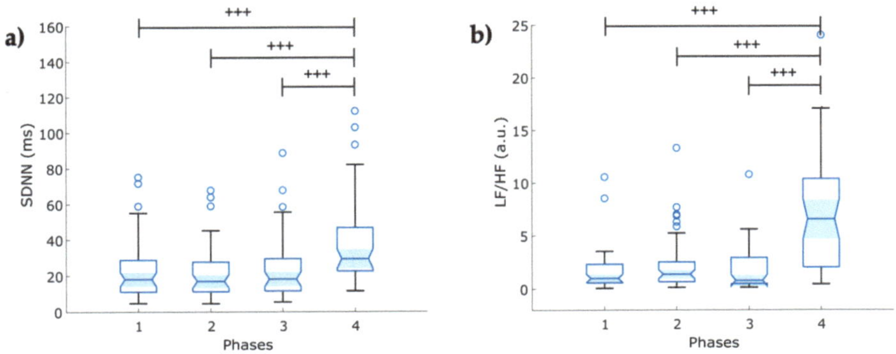

Figure 7. Boxplot of linear indices of the Pulse Rate variability (PRV) extracted from the photoplethysmographic signals (PPG) in COVID-19 survivors (N = 49) under a psychophysiological assessment of four phases: (**a**) mean values of the standard deviation of the IBI of typical sinus beats (SDNN); (**b**) ratio of low frequency (LF) and high frequency (HF) power. The figure shows ○ outliers, $^{+++}$ $p < 0.001$, obtained by Wilcoxon signed-rank test.

3.2.2. Nonlinear Features

The nonlinear features exhibited significantly decreased mean values for Phase 4: SampEn and a_2 long-term fluctuations decrease when breathing at a 0.1 Hz rate (Figure 8a,c) rather than following the other three breathing methods. However, short-term fluctuations (a_1) seemed to stand out in Phase 4 (Figure 8b) from the rest of the stages.

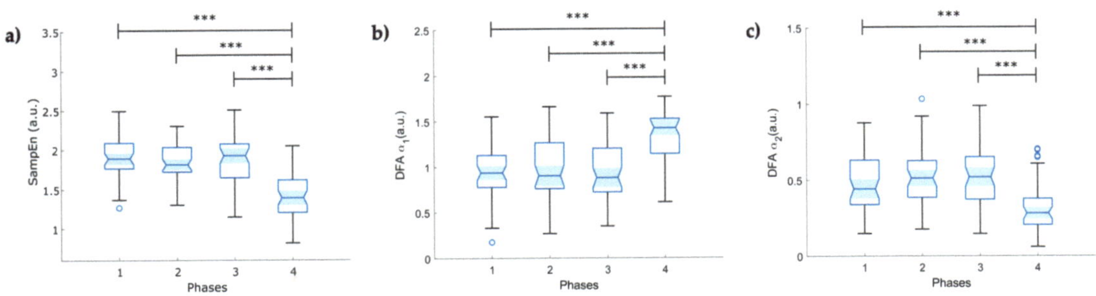

Figure 8. Boxplot of nonlinear indices of the pulse rate variability (PRV) extracted from the photoplethysmographic signals (PPG) in COVID-19 survivors (N = 49) under a psychophysiological assessment of four phases: (**a**) sample entropy (SampEn); (**b**,**c**) detrended fluctuation analysis between successive PP intervals for short (a_1) and long-term (a_2) fluctuations, respectively. The figure shows ○ outliers, *** $p < 0.001$, obtained by paired t-test.

3.3. Pulse–Respiration Quotient (PRQ) Variability

From the instantaneous PRQ time series, Phase 4 reported increased mean PRQ and SD-PRQ values compared to the rest of the stages ($p < 0.001$, Figure 9a,b). Concerning the nonlinear analysis, we observed a significant decrease in both entropy indexes ($p < 0.001$) at Phase 4 of the protocol (Figure 9c,d).

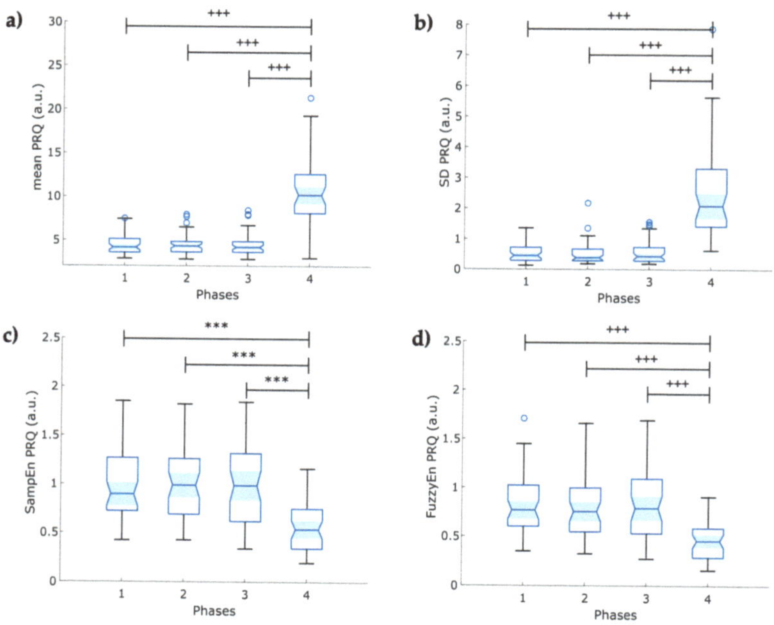

Figure 9. Boxplot of pulse–respiration quotient (PRQ) indexes obtained from the instantaneous heart rate and breathing rate time series (N = 49); (**a**) Mean PRQ, during slow breathing (6 breaths/min), a significant increase in PRQ is observed; (**b**) fluctuations of the instantaneous PRQ reported as the PRQ standard deviation; (**c**) Sample entropy of the instantaneous PRQ; (**d**) Fuzzy entropy of the PRQ. The figure shows ∘ outliers, *** $p < 0.001$ obtained by paired t-test, +++ $p < 0.001$ obtained by Wilcoxon signed-rank test.

The mean and standard deviation of all evaluated indexes from RESP (Table A1), PPG (Table A2), and PRQ (Table A3) time series are listed in Appendix A to show the main changes among protocol phases. The table also includes the p-values of the statistical test when comparing the first three phases with the last phase. All reported parameters showed a statistical significance with $p < 0.05$.

3.4. Comorbidity-Based Analysis

We reported the differences among comorbidity groups at the slow breathing phase. The features of these groups are listed in Table 4. We observed a mean heart rate difference with statistical significance ($p = 0.03$) between the no comorbidities (NOC) and diabetic (DIAB) groups (depicted in Figure 10a) but no significant variation with cardiovascular (CARD) and lung disease (PULM) groups.

Table 4. Clinical characteristics by comorbidity.

Comorbidity	n	Age (Years ± SD)
Cardiovascular disease	3	49 ± 14
Lung disease	3	53.75 ± 13.1
Diabetes	4	50.8 ± 13.8
No comorbidities	11	47.5 ± 13.1

Data presented in this table correspond to the group with n = 21.

Although the p-value obtained from a Kruskal–Wallis test for the mean PRQ did not show a significant difference, a trend ($p = 0.0748$) was observed among diabetic individuals compared to those with no pre-existing comorbidities (Figure 10b).

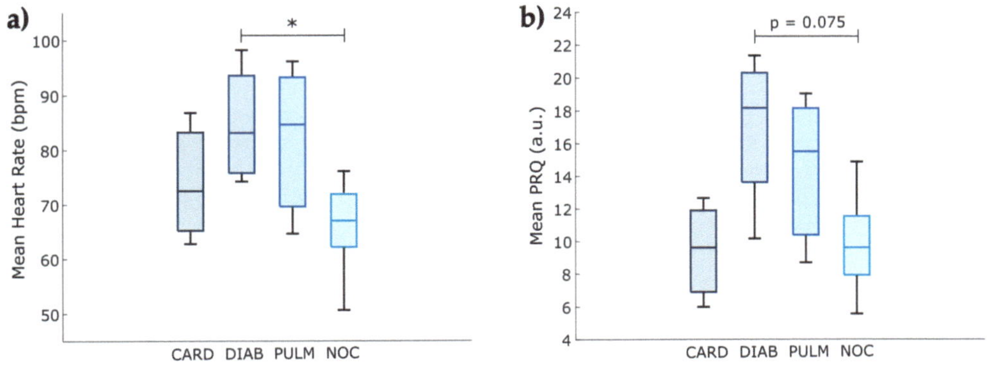

Figure 10. Boxplot of the parameters evaluated by individuals' comorbidities, comparing the Phase 4 obtained from the photoplethysmographic and pulse–respiration quotient time series of COVID-19 survivors (N = 21). (**a**) The mean heart rate changes among the comorbidity groups in the fourth phase of the test. (**b**) The mean PRQ index comparison among the cardiovascular disease (CARD), diabetes (DIAB), lung disease (PULM), and no comorbidities (NOC) groups in the fourth phase of the test. The figure displays bpm (beats per minute), with * indicating $p < 0.05$ obtained by the Kruskal–Wallis test.

4. Discussion

In this study, we analyzed whether the diaphragmatic slow-paced breathing technique improves the cardiorespiratory dynamics of COVID-19 survivors. In addition, we evaluated whether the patients' comorbidities significantly affect pulse and respiration rates.

4.1. Breathing Rate Variability

Firstly, BRV was studied to demonstrate that metrics obtained from RESP signals show differences among phases (Figure 6A). For Phases 1 to 3, we found lower values in SDBB and RMSSD-BB (less than 1 s). This finding is consistent with Alhuthail et al. [51], who found RMSSD and SDBB values less than 1 s in post-COVID subjects using structured light plethysmography. Additionally, we confirmed that the psychophysiological assessment modified the mean breathing rate in Phase 4 compared to the other three phases. Moreover, we found that SDBB (Figure 6b) and RMSSD (Figure 6c) indices increased in Phase 4 compared to the previous phases; both indices are indicators of the total variability of breathing and the short-term variability between breathing cycles, respectively [38]. Research on BRV in COVID-19 survivors is scarce, and even less is known about the impact of diaphragmatic breathing in this population. Particularly, relevant research has been published recently about the impact of meditation on breathing rate and BRV [28,52,53]. Some authors have reported increased RMSSD related to breathing awareness in meditator and non-meditator participants [38]. This observation may partially explain our findings.

4.2. Pulse Rate Variability

From the linear analysis of PRV, we found that slow-paced breathing (Phase 4) significantly increases cardiac parasympathetic activity and global variability, depicted by SDNN and RMSSD-PP values (Table A2, Appendix A). This finding is consistent with You et al. [54], who found an increase in the RMSSD during slow-paced breathing at six breaths per minute compared to the baseline condition in healthy subjects. In the frequency analysis, we found that slow-paced breathing increased the LF power band and decreased the HF power band in Phase 4 compared to previous phases (Table A2). These findings are consistent with the findings of Li et al. [55] on uncorrected spectral HRV analysis, which showed lower HF power, increased LF band, and LF/HF ratio in patients with hypertension under 8 breaths/min breathing. In contrast, the study published by Sakakibara [25] found increased HRV during paced breathing at eight breaths per minute in healthy individuals with an increased HF power band. A possible explanation for this might be that 43% of our participants were COVID-19 survivors with no comorbidities, while in the study by Li et al., the pattern was shown in the group with hypertension. Hypertension reduces the activity of the parasympathetic branch of the ANS and increases sympathetic activity [56,57]. Moreover, the observed decrease in the HF power band in our group in Phase 4 could be partially explained by the autonomic imbalance reported in COVID-19 survivors.

In the nonlinear analysis, we observed a decrease in SampEn during slow-paced breathing. This parameter measures the regularity and complexity of a time series, with lower values indicating more regular patterns [43]. In our group, this index provides evidence that slow-paced breathing alters the regulation of the cardiac system. We did not recruit a control group comprising healthy subjects to evaluate possible autonomic impairment. However, other authors have reported that COVID-19 survivors exhibit lower SampEn values, which reflect a higher degree of autonomic impairment, compared to healthy control groups [58,59]. In this study, the SampEn values are higher than those reported previously. A possible explanation for this difference might be due to the inclusion of participants with diabetes and cardiovascular diseases (as shown in Tables 1 and 4). Bajić et al. [59] found that patients with diabetes and hypertension exhibited higher sample entropy among the comorbidities listed.

Previous studies have reported a distinct behavior of the DFA scaling exponents in COVID-19 survivors compared to a healthy group. Our results partially agree with those reported by Kurtoğlu et al. [58]. Our group demonstrated that the long-term scaling exponent α_2 value was lower than theirs for all phases. This difference can be partially explained by our small sample size, compared to other studies, and the comorbidities present in our group. Additionally, a study concluded that diabetic individuals exhibited an increased α_2 compared to the short-term exponent α_1, which indicates decreased parasympathetic activity. Conversely, healthy subjects showed the opposite [60].

Referring to Phase 4, we observed that slow-paced breathing modifies the DFA exponents, with an increase in the short-term exponent α_1 and a decrease in the long-term exponent α_2 (Figure 8b,c), which may be related to the reciprocal regulation between the cardiac and respiratory systems, influenced by slow breathing or an individual's posture, which enhances the parasympathetic influence on cardiac function. Matić et al. found that short-term scaling α_1 increased significantly, while the long-term scaling exponent α_2 decreased in standing and during slow breathing, indicating reciprocal regulation. Their study showed that changes in posture and breathing patterns significantly impact cardiac and respiratory interactions mediated through sympathovagal control [61]. Our results demonstrated that nonlinear indices (SampEn and DFA) had lower p-values than linear indices (Table A2, Appendix A).

4.3. Pulse–Respiration Quotient

The PRQ analysis reveals complex cardiorespiratory interactions that are not perceptible through the single analysis of BRV and PRV [22]. Currently, studies related to PRQ primarily involve healthy individuals, and there is no data regarding COVID-19 survivors who experienced post-acute syndrome or performed slow-paced breathing. Matić et al. reported increased mean PRQ and PRQ variability with orthostasis and slow-paced breathing as signs of cardiorespiratory coupling regulation and adaptability [62]. Our findings on the mean PRQ (Figure 9a) and SD-PRQ (Figure 9b) showed a significant difference in response to the change in breathing frequency with higher statistical significance than the mean and SD of PRV indices. However, contrary to this study, our mean PRQ and SD PRQ indices exhibited a slightly lower p-value than SD RESP (Table A3).

We found that lower values of SampEn (Figure 9c) and FuzzyEn (Figure 9d) of the PRQ time series might indicate higher cardiorespiratory coupling mediated by the ANS in the induced-relaxation phase, as suggested by Tian and Song, who reported a decrease in the FuzzyEn index and BR during slow breathing in healthy participants [63].

4.4. Comorbidity-Based Analysis

Related to the effects of slow breathing on COVID-19 survivors with a previously detected comorbidity, we found that even though changes that enhance autonomic regulation were visible when performing diaphragmatic breathing, patients with type 2 diabetes mellitus (T2DM) still showed a higher heart rate (Figure 10a), possibly associated with cardiac autonomic neuropathy [64,65]. Benichou et al. reported that T2DM patients showed a decreased value of all HRV parameters, which exhibited restricted sympathetic and parasympathetic activity [66]. Another study demonstrated a short-term reduction of stress levels, enhanced glycemic status (reduced hyperglycemia), and increased immune response in diabetic patients with COVID-19 who followed a relaxation technique [67].

Vanzella et al. [68] assessed the changes in HRV among healthy individuals and patients with COPD; a significant decrease in time (RMSSD, SDNN) and frequency (LF, HF) domain indices of patients with COPD were found, as indicators of the reduction of both sympathetic and parasympathetic activity that suggests a worsened autonomic modulation. The revision conducted by Hamasaki concluded that diaphragmatic breathing improved the respiratory function of patients with COPD, reduced stress and anxiety symptoms by stimulating vagal activation, and enhanced cardiorespiratory performance in patients with heart failure [69].

Lachowska et al. reported that HR and blood pressure responses decreased considerably with slow breathing exercises in heart failure patients with reduced ejection fraction and enhanced a variety of health-related scores of quality of life (QoL) [70], while another study exhibited improvement in the long-term cardiorespiratory capacity of these patients [71]. Post-stroke patients also seem to benefit from slow breathing, with increased baroreflex sensitivity, lower systolic blood pressure, and HR with increased HRV [72].

Apart from the diabetic group, no significant difference was found among comorbidities and healthy individuals during slow breathing (Figure 10a,b). We supposed that these results could be associated with the fact that both conditions (comorbidities and post-COVID-19 syndrome) have been proven to modify autonomic function, and the reduced sample size may cause no statistically significant difference among comorbidity groups.

4.5. Limitations and Future Work

In this study, we encountered some limitations; first, the reduced sample size to address autonomic changes among patients with comorbidities and healthy individuals, where both groups coursed COVID-19 sequelae, and the lack of previous research on the use of the PRQ to evaluate cardiac and respiratory interactions in a similar population to the one presented in this work, that must be considered in the interpretation of the provided data. Despite participants reducing their BR, we also noted that several did not meet the 6 breaths/min breathing rate (Figure 6a and Table A1) that can be partly associated with the group comorbidities and symptoms of post-COVID-19 syndrome. Another possible explanation is that many participants did not have previous training in diaphragmatic breathing.

COVID-19 sequelae negatively impact survivors' physical and mental health in the short and long term [73–75]. It has been observed that most of these patients exhibited high levels of cellular inflammation and autonomic dysregulation [76]. Slow breathing practice has been proven to enhance the autonomic regulation of individuals with diverse chronic illnesses [77]; it improves cardiovascular response, decreases sympathetic hyperactivation, and induces a state of physiological balance, which may be a critical factor in the prevention or treatment of the cardiovascular manifestations [78–80]. Additionally, it reduces anxious and depressive symptoms, which can worsen the state of health and recovery of these patients [81,82]. This psychophysiological evaluation allows observing the response to various conditions to identify adaptive and maladaptive response patterns to intervene and, if necessary, modify them [31,81].

Additional studies may be conducted to assess the long-term effect of slow breathing and other psychophysiological interventions on COVID-19 survivors who present post-infection sequelae and may or may not have other comorbidities. Moreover, further analysis of the differences between men and women is suggested with an increased sample size. However, the study of the PRQ index seems like a feasible technique to evaluate cardiac and respiratory interactions and the changes mediated by breathing rate modifications.

5. Conclusions

Our results demonstrate that implementing a psychophysiological assessment based on slow paced-breathing induces significant changes in the breathing rate variability and pulse rate variability of COVID-19 survivors. These changes may improve the cardiorespiratory dynamics of this population by enhancing cardiorespiratory coupling in the short-term via increased vagal activity. Additionally, our findings suggest that patients with comorbidities, such as type 2 diabetes mellitus, may still benefit from diaphragmatic breathing despite changes in sympathovagal balance. However, further research is needed to confirm the efficacy of this practice in the long term.

Author Contributions: Conceptualization, V.P.-H., G.L.L.-R. and A.O.-T.; data curation, A.M.S.-S.; formal analysis, A.M.S.-S.; funding acquisition, V.P.-H.; investigation, V.P.-H., G.L.L.-R. and A.O.-T.; methodology, A.M.S.-S., L.M.S.-F. and J.J.R.-L.; project administration, V.P.-H., L.M.S.-F., G.L.L.-R. and J.J.R.-L.; resources, V.P.-H.; software, A.M.S.-S.; supervision, V.P.-H., L.M.S.-F., G.L.L.-R. and J.J.R.-L.; validation, A.M.S.-S.; visualization, A.M.S.-S.; writing—original draft, A.M.S.-S., V.P.-H., L.M.S.-F., G.L.L.-R., J.J.R.-L. and A.O.-T.; writing—review and editing, A.M.S.-S., V.P.-H., L.M.S.-F., G.L.L.-R., J.J.R.-L. and A.O.-T. All authors have read and agreed to the published version of the manuscript.

Funding: This study was funded in part by Fundación Gonzalo Río Arronte and Instituto Nacional de Enfermedades Respiratorias Ismael Cosío Villegas. The study sponsors had no role in the data collection, analysis, or interpretation of results.

Institutional Review Board Statement: The study was conducted in accordance with the Declaration of Helsinki and approved by the Ethics Committee of the National Institute of Respiratory Diseases, Mexico (approval number C57-21, 13 December 2021).

Data Availability Statement: The access to the data presented in this study should be requested of the authors and may be restricted due to hospital confidentiality policy.

Conflicts of Interest: The authors declare no conflict of interest.

Appendix A

Table A1. Mean values (±standard deviation) of indexes obtained from breathing signals at four phases of the protocol. Statistical significance from paired *t*-test and Wilcoxon signed-rank test are included.

Index	Phase 1	Phase 2	Phase 3	Phase 4	Comparison	*p*
Mean BR	18.398 ± 4.16	18.498 ± 3.81	18.249 ± 4.13	9.303 ± 3.59		
					P1–P4	2.21×10^{-18} *
					P2–P4	5.69×10^{-20} *
					P3–P4	2.66×10^{-18} *
SDBB	0.424 ± 0.25	0.395 ± 0.26	0.450 ± 0.31	2.077 ± 1.32		
					P1–P4	2.69×10^{-11} +
					P2–P4	1.85×10^{-11} +
					P3–P4	1.60×10^{-10} +
RMSSD	0.436 ± 0.25	0.417 ± 0.28	0.436 ± 0.27	2.488 ± 1.99		
					P1–P4	4.74×10^{-9} +
					P2–P4	3.45×10^{-9} +
					P3–P4	3.97×10^{-9} +

BR—breathing rate; SDBB—standard deviation of breath to breath (BB) intervals; RMSSD—root mean square of successive BB interval differences; P1—Phase 1; P2—Phase 2; P3—Phase 3; P4—Phase 4. * *p*-value obtained by paired *t*-test, + *p*-value obtained by Wilcoxon signed-rank test.

Table A2. Mean values (±standard deviation) of linear and nonlinear indexes obtained from photoplethysmographic signals at four phases of the protocol. Statistical significance from paired *t*-test and Wilcoxon signed-rank test are included.

Index	Phase 1	Phase 2	Phase 3	Phase 4	Comparison	p
Mean PP	820.60 ± 129.22	808.54 ± 125.42	821.75 ± 135.57	840.91 ± 137.29		
					P1–P4	2.59×10^{-4} *
					P2–P4	3.29×10^{-7} *
					P3–P4	3.48×10^{-4} *
SDNN	22.49 ± 15.82	21.23 ± 14.79	23.56 ± 17.25	37.99 ± 23.76		
					P1–P4	4.52×10^{-9} +
					P2–P4	3.55×10^{-9} +
					P3–P4	2.47×10^{-8} +
RMSSD	23.90 ± 21.15	21.99 ± 19.25	24.35 ± 22.74	28.25 ± 20.23		
					P1–P4	0.0013 +
					P2–P4	1.85×10^{-6} +
					P3–P4	3.91×10^{-4} +
LF power (nu)	42.63 ± 17.71	47.80 ± 19.87	44.25 ± 20.18	73.85 ± 19.47		
					P1–P4	7.19×10^{-8} +
					P2–P4	2.25×10^{-6} +
					P3–P4	6.74×10^{-7} +
HF power (nu)	44.29 ± 22.10	39.04 ± 21.26	43.98 ± 22.58	20.28 ± 16.81		
					P1–P4	6.98×10^{-9} *
					P2–P4	5.24×10^{-6} *
					P3–P4	5.16×10^{-7} *
LF/HF ratio	1.65 ± 1.92	2.21 ± 2.52	1.88 ± 2.12	7.06 ± 5.37		
					P1–P4	5.75×10^{-8} +
					P2–P4	2.14×10^{-6} +
					P3–P4	2.08×10^{-7} +
SampEn	1.91 ± 0.27	1.84 ± 0.24	1.86 ± 0.28	1.41 ± 0.31		
					P1–P4	2.83×10^{-11} *
					P2–P4	4.42×10^{-11} *
					P3–P4	5.93×10^{-10} *
DFA α_1	0.95 ± 0.29	0.96 ± 0.32	0.96 ± 0.32	1.33 ± 0.29		
					P1–P4	9.30×10^{-10} *
					P2–P4	4.11×10^{-9} *
					P3–P4	6.63×10^{-9} *
DFA α_2	0.47 ± 0.18	0.52 ± 0.2	0.51 ± 0.19	0.32 ± 0.17		
					P1–P4	2.60×10^{-6} *
					P2–P4	1.83×10^{-7} *
					P3–P4	6.85×10^{-7} *
Mean HR	75.00 ± 11.47	76.04 ± 11.34	75.00 ± 11.6	73.44 ± 11.55		
					P1–P4	6.27×10^{-4} *
					P2–P4	1.89×10^{-6} *
					P3–P4	9.10×10^{-4} *

PP—peak-to-peak; SDNN—standard deviation of normal PP intervals; RMSSD—root mean square of successive RR interval differences; LF—relative power of the low-frequency band; HF—relative power of the high-frequency band; n.u.—normalized unit; LF/HF—ratio of low to high-frequency band power; SampEn—sample entropy; DFA α_1—detrended fluctuation analysis (short-term); DFA α_2—detrended fluctuation analysis (long-term); HR—heart rate; SD—standard deviation; P1—Phase 1; P2—Phase 2; P3—Phase 3; P4—Phase 4. * p-value obtained by paired *t*-test, + p-value obtained by Wilcoxon signed-rank test.

Table A3. Mean values (±standard deviation) of indexes obtained from instantaneous pulse–respiration quotient time series at four phases of the protocol. Statistical significance from paired t-test and Wilcoxon signed-rank test are included.

Index	Phase 1	Phase 2	Phase 3	Phase 4	Comparison	p
Mean PRQ	4.37 ± 1.102	4.38 ± 1.12	4.44 ± 1.267	10.83 ± 3.875		
					P1–P4	1.34×10^{-9} +
					P2–P4	1.18×10^{-9} +
					P3–P4	1.34×10^{-9} +
SD—PRQ	0.538 ± 0.30	0.535 ± 0.372	0.583 ± 0.399	2.49 ± 1.523		
					P1–P4	1.18×10^{-9} +
					P2–P4	1.11×10^{-9} +
					P3–P4	2.19×10^{-9} +
SampEn	0.973 ± 0.348	0.99 ± 0.36	0.995 ± 0.419	0.563 ± 0.24		
					P1–P4	5.70×10^{-9} *
					P2–P4	3.10×10^{-9} *
					P3–P4	1.21×10^{-7} *
FuzzyEn	0.824 ± 0.309	0.825 ± 0.315	0.821 ± 0.352	0.471 ± 0.205		
					P1–P4	3.10×10^{-8} +
					P2–P4	1.46×10^{-7} +
					P3–P4	2.60×10^{-6} +

PRQ—pulse–respiration quotient; SD—standard deviation; SampEn—sample entropy; FuzzyEn—fuzzy entropy; P1—Phase 1; P2—Phase 2; P3—Phase 3; P4—Phase 4. * p-value obtained by paired t-test, + p-value obtained by Wilcoxon signed-rank test.

References

1. Salian, V.S.; Wright, J.A.; Vedell, P.T.; Nair, S.; Li, C.; Kandimalla, M.; Tang, X.; Carmona Porquera, E.M.; Kalari, K.R.; Kandimalla, K.K. COVID-19 Transmission, Current Treatment, and Future Therapeutic Strategies. *Mol. Pharm.* **2021**, *18*, 754–771. [CrossRef] [PubMed]
2. Ahmad Malik, J.; Ahmed, S.; Shinde, M.; Almermesh, M.H.S.; Alghamdi, S.; Hussain, A.; Anwar, S. The Impact of COVID-19 On Comorbidities: A Review Of Recent Updates For Combating It. *Saudi J. Biol. Sci.* **2022**, *29*, 3586–3599. [CrossRef] [PubMed]
3. Dong, E.; Du, H.; Gardner, L. An Interactive Web-Based Dashboard to Track COVID-19 in Real Time. *Lancet Infect. Dis.* **2020**, *20*, 533–534. [CrossRef] [PubMed]
4. Anka, A.U.; Tahir, M.I.; Abubakar, S.D.; Alsabbagh, M.; Zian, Z.; Hamedifar, H.; Sabzevari, A.; Azizi, G. Coronavirus Disease 2019 (COVID-19): An Overview of the Immunopathology, Serological Diagnosis and Management. *Scand. J. Immunol.* **2021**, *93*, e12998. [CrossRef]
5. Elrobaa, I.H.; New, K.J. COVID-19: Pulmonary and Extra Pulmonary Manifestations. *Front. Public Health* **2021**, *9*, 711616. [CrossRef] [PubMed]
6. Long, B.; Carius, B.M.; Chavez, S.; Liang, S.Y.; Brady, W.J.; Koyfman, A.; Gottlieb, M. Clinical Update on COVID-19 for the Emergency Clinician: Presentation and Evaluation. *Am. J. Emerg. Med.* **2022**, *54*, 46–57. [CrossRef]
7. *COVID-19 Rapid Guideline: Managing the Long-Term Effects of COVID-19*; National Institute for Health and Care Excellence: Clinical Guidelines; National Institute for Health and Care Excellence (NICE): London, UK, 2020; ISBN 978-1-4731-3943-5.
8. Logue, J.K.; Franko, N.M.; McCulloch, D.J.; McDonald, D.; Magedson, A.; Wolf, C.R.; Chu, H.Y. Sequelae in Adults at 6 Months After COVID-19 Infection. *JAMA Netw. Open* **2021**, *4*, e210830. [CrossRef]
9. Munker, D.; Veit, T.; Barton, J.; Mertsch, P.; Mümmler, C.; Osterman, A.; Khatamzas, E.; Barnikel, M.; Hellmuth, J.C.; Münchhoff, M.; et al. Pulmonary Function Impairment of Asymptomatic and Persistently Symptomatic Patients 4 Months after COVID-19 According to Disease Severity. *Infection* **2022**, *50*, 157–168. [CrossRef]
10. Raman, B.; Bluemke, D.A.; Lüscher, T.F.; Neubauer, S. Long COVID: Post-Acute Sequelae of COVID-19 with a Cardiovascular Focus. *Eur. Heart J.* **2022**, *43*, 1157–1172. [CrossRef]
11. Desai, A.D.; Lavelle, M.; Boursiquot, B.C.; Wan, E.Y. Long-Term Complications of COVID-19. *Am. J. Physiol.-Cell Physiol.* **2022**, *322*, C1–C11. [CrossRef]
12. Castanares-Zapatero, D.; Chalon, P.; Kohn, L.; Dauvrin, M.; Detollenaere, J.; Maertens de Noordhout, C.; Primus-de Jong, C.; Cleemput, I.; Van den Heede, K. Pathophysiology and Mechanism of Long COVID: A Comprehensive Review. *Ann. Med.* **2022**, *54*, 1473–1487. [CrossRef] [PubMed]
13. Becker, R.C. Autonomic Dysfunction in SARS-CoV-2 Infection Acute and Long-Term Implications COVID-19 Editor's Page Series. *J. Thromb. Thrombolysis* **2021**, *52*, 692–707. [CrossRef] [PubMed]
14. Al-kuraishy, H.M.; Al-Gareeb, A.I.; Qusti, S.; Alshammari, E.M.; Gyebi, G.A.; Batiha, G.E.-S. COVID-19-Induced Dysautonomia: A Menace of Sympathetic Storm. *ASN Neuro* **2021**, *13*, 175909142110576. [CrossRef] [PubMed]
15. Lopez-Leon, S.; Wegman-Ostrosky, T.; Perelman, C.; Sepulveda, R.; Rebolledo, P.A.; Cuapio, A.; Villapol, S. More than 50 Long-Term Effects of COVID-19: A Systematic Review and Meta-Analysis. *Sci. Rep.* **2021**, *11*, 16144. [CrossRef]

16. Barizien, N.; Le Guen, M.; Russel, S.; Touche, P.; Huang, F.; Vallée, A. Clinical Characterization of Dysautonomia in Long COVID-19 Patients. *Sci. Rep.* **2021**, *11*, 14042. [CrossRef]
17. Dotan, A.; David, P.; Arnheim, D.; Shoenfeld, Y. The Autonomic Aspects of the Post-COVID19 Syndrome. *Autoimmun. Rev.* **2022**, *21*, 103071. [CrossRef]
18. Ser, M.H.; Çalıkuşu, F.Z.; Tanrıverdi, U.; Abbaszade, H.; Hakyemez, S.; Balkan, İ.İ.; Karaali, R.; Gündüz, A. Autonomic and Neuropathic Complaints of Long-COVID Objectified: An Investigation from Electrophysiological Perspective. *Neurol. Sci.* **2022**, *43*, 6167–6177. [CrossRef]
19. Charleston-Villalobos, S.; Reulecke, S.; Voss, A.; Azimi-Sadjadi, M.R.; González-Camarena, R.; Gaitán-González, M.J.; González-Hermosillo, J.A.; Hernández-Pacheco, G.; Schulz, S.; Aljama-Corrales, T. Time-Frequency Analysis of Cardiovascular and Cardiorespiratory Interactions During Orthostatic Stress by Extended Partial Directed Coherence. *Entropy* **2019**, *21*, 468. [CrossRef]
20. Santiago-Fuentes, L.M.; Charleston-Villalobos, S.; González-Camarena, R.; Voss, A.; Mejía-Avila, M.E.; Buendía-Roldan, I.; Reulecke, S.; Aljama-Corrales, T. Effects of Supplemental Oxygen on Cardiovascular and Respiratory Interactions by Extended Partial Directed Coherence in Idiopathic Pulmonary Fibrosis. *Front. Netw. Physiol.* **2022**, *2*, 834056. [CrossRef]
21. Thayer, J.F.; Yamamoto, S.S.; Brosschot, J.F. The Relationship of Autonomic Imbalance, Heart Rate Variability and Cardiovascular Disease Risk Factors. *Int. J. Cardiol.* **2010**, *141*, 122–131. [CrossRef]
22. Scholkmann, F.; Wolf, U. The Pulse-Respiration Quotient: A Powerful but Untapped Parameter for Modern Studies About Human Physiology and Pathophysiology. *Front. Physiol.* **2019**, *10*, 371. [CrossRef] [PubMed]
23. Rodríguez Medina, D.A. Efectos Psicofisiológicos de La Respiración Diafragmática y La Estimulación Térmica Sobre La Actividad Autonómica Del Estrés Agudo. *Acta Investig. Psicológica* **2018**, *8*, 101–113. [CrossRef]
24. Hopper, S.I.; Murray, S.L.; Ferrara, L.R.; Singleton, J.K. Effectiveness of Diaphragmatic Breathing on Physiological and Psychological Stress in Adults: A Quantitative Systematic Review Protocol. *JBI Database Syst. Rev. Implement. Rep.* **2018**, *16*, 1367–1372. [CrossRef]
25. Sakakibara, M. Evaluation of Heart Rate Variability and Application of Heart Rate Variability Biofeedback: Toward Further Research on Slow-Paced Abdominal Breathing in Zen Meditation. *Appl. Psychophysiol. Biofeedback* **2022**, *47*, 345–356. [CrossRef] [PubMed]
26. Anasuya, B.; Deepak, K.; Jaryal, A.; Narang, R. Effect of Slow Breathing on Autonomic Tone & Baroreflex Sensitivity in Yoga Practitioners. *Indian J. Med. Res.* **2020**, *152*, 638. [CrossRef] [PubMed]
27. Ma, X.; Yue, Z.-Q.; Gong, Z.-Q.; Zhang, H.; Duan, N.-Y.; Shi, Y.-T.; Wei, G.-X.; Li, Y.-F. The Effect of Diaphragmatic Breathing on Attention, Negative Affect and Stress in Healthy Adults. *Front. Psychol.* **2017**, *8*, 874. [CrossRef] [PubMed]
28. Behan, C. The Benefits of Meditation and Mindfulness Practices during Times of Crisis Such as COVID-19. *Ir. J. Psychol. Med.* **2020**, *37*, 256–258. [CrossRef] [PubMed]
29. Sevoz-Couche, C.; Laborde, S. Heart Rate Variability and Slow-Paced Breathing:When Coherence Meets Resonance. *Neurosci. Biobehav. Rev.* **2022**, *135*, 104576. [CrossRef]
30. Laborde, S.; Allen, M.S.; Borges, U.; Dosseville, F.; Hosang, T.J.; Iskra, M.; Mosley, E.; Salvotti, C.; Spolverato, L.; Zammit, N.; et al. Effects of Voluntary Slow Breathing on Heart Rate and Heart Rate Variability: A Systematic Review and a Meta-Analysis. *Neurosci. Biobehav. Rev.* **2022**, *138*, 104711. [CrossRef]
31. Corrado, J.; Halpin, S.; Preston, N.; Whiteside, D.; Tarrant, R.; Davison, J.; Simms, A.D.; O'Connor, R.J.; Casson, A.; Sivan, M. HEART Rate Variability Biofeedback for Long COVID Symptoms (HEARTLOC): Protocol for a Feasibility Study. *BMJ Open* **2022**, *12*, e066044. [CrossRef]
32. Peláez-Hernández, V.; Luna-Rodríguez, G.L.; Orea-Tejeda, A.; Mora-Gallegos, J.; Keirns-Davis, C.; González-Islas, D. Heart Rate Variability Disturbances and Biofeedback Treatment in COVID-19 Survivors. *E-J. Cardiol. Pract.* **2021**, *21*. Available online: https://www.escardio.org/Journals/E-Journal-of-Cardiology-Practice/Volume-21/heart-rate-variability-disturbances-and-biofeedback-treatment-in-covid-19-surviv (accessed on 30 March 2023).
33. Domínguez Trejo, B.; Ruvalcaba Palacios, G.; Montero López-Lena, M. Pain, Emotions, and Social-Well-Being in Mexico. In *Handbook of Happiness Research in Latin America*; Rojas, M., Ed.; Springer: Dordrecht, The Netherlands, 2016; pp. 489–513. ISBN 978-94-017-7202-0.
34. van Goor, H.M.R.; van Loon, K.; Breteler, M.J.M.; Kalkman, C.J.; Kaasjager, K.A.H. Circadian Patterns of Heart Rate, Respiratory Rate and Skin Temperature in Hospitalized COVID-19 Patients. *PLoS ONE* **2022**, *17*, e0268065. [CrossRef] [PubMed]
35. Spengler, C.M.; Czeisler, C.A.; Shea, S.A. An Endogenous Circadian Rhythm of Respiratory Control in Humans. *J. Physiol.* **2000**, *526*, 683–694. [CrossRef] [PubMed]
36. Voss, A.; Schroeder, R.; Heitmann, A.; Peters, A.; Perz, S. Short-Term Heart Rate Variability—Influence of Gender and Age in Healthy Subjects. *PLoS ONE* **2015**, *10*, e0118308. [CrossRef]
37. The MathWorks Inc. Signal Processing Toolbox 2022. Available online: https://www.mathworks.com/products/signal.html (accessed on 7 February 2023).
38. Soni, R.; Muniyandi, M. Breath Rate Variability: A Novel Measure to Study the Meditation Effects. *Int. J. Yoga* **2019**, *12*, 45. [CrossRef] [PubMed]
39. Mejía-Mejía, E.; May, J.M.; Torres, R.; Kyriacou, P.A. Pulse Rate Variability in Cardiovascular Health: A Review on Its Applications and Relationship with Heart Rate Variability. *Physiol. Meas.* **2020**, *41*, 07TR01. [CrossRef]
40. Yuda, E.; Shibata, M.; Ogata, Y.; Ueda, N.; Yambe, T.; Yoshizawa, M.; Hayano, J. Pulse Rate Variability: A New Biomarker, Not a Surrogate for Heart Rate Variability. *J. Physiol. Anthropol.* **2020**, *39*, 21. [CrossRef]

41. Tarvainen, M.P.; Niskanen, J.-P.; Lipponen, J.A.; Ranta-aho, P.O.; Karjalainen, P.A. Kubios HRV—Heart Rate Variability Analysis Software. *Comput. Methods Programs Biomed.* **2014**, *113*, 210–220. [CrossRef]
42. Shaffer, F.; Ginsberg, J.P. An Overview of Heart Rate Variability Metrics and Norms. *Front. Public Health* **2017**, *5*, 258. [CrossRef]
43. Richman, J.S.; Moorman, J.R. Physiological Time-Series Analysis Using Approximate Entropy and Sample Entropy. *Am. J. Physiol.-Heart Circ. Physiol.* **2000**, *278*, H2039–H2049. [CrossRef]
44. Solís-Montufar, E.E.; Gálvez-Coyt, G.; Muñoz-Diosdado, A. Entropy Analysis of RR-Time Series From Stress Tests. *Front. Physiol.* **2020**, *11*, 981. [CrossRef]
45. Hansen, C.; Wei, Q.; Shieh, J.-S.; Fourcade, P.; Isableu, B.; Majed, L. Sample Entropy, Univariate, and Multivariate Multi-Scale Entropy in Comparison with Classical Postural Sway Parameters in Young Healthy Adults. *Front. Hum. Neurosci.* **2017**, *11*, 206. [CrossRef] [PubMed]
46. Kamath, M.V.; Watanabe, M.; Upton, A. *Heart Rate Variability (HRV) Signal Analysis: Clinical Applications*, 1st ed.; Chapman and Hall/CRC: Baton Rouge, LA, USA, 2016; ISBN 978-1-4665-7605-6.
47. Chen, W.; Wang, Z.; Xie, H.; Yu, W. Characterization of Surface EMG Signal Based on Fuzzy Entropy. *IEEE Trans. Neural Syst. Rehabil. Eng.* **2007**, *15*, 266–272. [CrossRef] [PubMed]
48. De Luca, A.; Termini, S. A Definition of a Nonprobabilistic Entropy in the Setting of Fuzzy Sets Theory. *Inf. Control* **1972**, *20*, 301–312. [CrossRef]
49. Chen, W.; Zhuang, J.; Yu, W.; Wang, Z. Measuring Complexity Using FuzzyEn, ApEn, and SampEn. *Med. Eng. Phys.* **2009**, *31*, 61–68. [CrossRef] [PubMed]
50. Silva, L.E.V.; Fazan, R.; Marin-Neto, J.A. PyBioS: A Freeware Computer Software for Analysis of Cardiovascular Signals. *Comput. Methods Programs Biomed.* **2020**, *197*, 105718. [CrossRef] [PubMed]
51. Alhuthail, E.; Stockley, J.; Coney, A.; Cooper, B. Measurement of Breathing in Patients with Post-COVID-19 Using Structured Light Plethysmography (SLP). *BMJ Open Respir. Res.* **2021**, *8*, e001070. [CrossRef]
52. Gerritsen, R.J.S.; Band, G.P.H. Breath of Life: The Respiratory Vagal Stimulation Model of Contemplative Activity. *Front. Hum. Neurosci.* **2018**, *12*, 397. [CrossRef]
53. Zaccaro, A.; Piarulli, A.; Laurino, M.; Garbella, E.; Menicucci, D.; Neri, B.; Gemignani, A. How Breath-Control Can Change Your Life: A Systematic Review on Psychophysiological Correlates of Slow Breathing. *Front. Hum. Neurosci.* **2018**, *12*, 353. [CrossRef]
54. You, M.; Laborde, S.; Zammit, N.; Iskra, M.; Borges, U.; Dosseville, F. Single Slow-Paced Breathing Session at Six Cycles per Minute: Investigation of Dose-Response Relationship on Cardiac Vagal Activity. *Int. J. Environ. Res. Public. Health* **2021**, *18*, 12478. [CrossRef]
55. Li, C.; Chang, Q.; Zhang, J.; Chai, W. Effects of Slow Breathing Rate on Heart Rate Variability and Arterial Baroreflex Sensitivity in Essential Hypertension. *Medicine* **2018**, *97*, e0639. [CrossRef]
56. Shanks, J.; Ramchandra, R. Angiotensin II and the Cardiac Parasympathetic Nervous System in Hypertension. *Int. J. Mol. Sci.* **2021**, *22*, 12305. [CrossRef]
57. Li, Y.; Wei, B.; Liu, X.; Shen, X.Z.; Shi, P. Microglia, Autonomic Nervous System, Immunity and Hypertension: Is There a Link? *Pharmacol. Res.* **2020**, *155*, 104451. [CrossRef] [PubMed]
58. Kurtoğlu, E.; Afsin, A.; Aktaş, İ.; Aktürk, E.; Kutlusoy, E.; Çağaşar, Ö. Altered Cardiac Autonomic Function after Recovery from COVID-19. *Ann. Noninvasive Electrocardiol.* **2022**, *27*, e12916. [CrossRef] [PubMed]
59. Bajić, D.; Đajić, V.; Milovanović, B. Entropy Analysis of COVID-19 Cardiovascular Signals. *Entropy* **2021**, *23*, 87. [CrossRef]
60. Roy, B.; Ghatak, S. Nonlinear Methods to Assess Changes in Heart Rate Variability in Type 2 Diabetic Patients. *Arq. Bras. Cardiol.* **2013**, *101*, 317–327. [CrossRef] [PubMed]
61. Matić, Z.; Platiša, M.M.; Kalauzi, A.; Bojić, T. Slow 0.1 Hz Breathing and Body Posture Induced Perturbations of RRI and Respiratory Signal Complexity and Cardiorespiratory Coupling. *Front. Physiol.* **2020**, *11*, 24. [CrossRef]
62. Matić, Z.; Kalauzi, A.; Moser, M.; Platiša, M.M.; Lazarević, M.; Bojić, T. Pulse Respiration Quotient as a Measure Sensitive to Changes in Dynamic Behavior of Cardiorespiratory Coupling Such as Body Posture and Breathing Regime. *Front. Physiol.* **2022**, *13*, 946613. [CrossRef]
63. Tian, N.; Song, R. Effects of Different Interventions on Cardiac Regulation Using Fuzzy Entropy. *IEEE Access* **2019**, *7*, 75949–75956. [CrossRef]
64. Vinik, A.I.; Casellini, C.; Parson, H.K.; Colberg, S.R.; Nevoret, M.-L. Cardiac Autonomic Neuropathy in Diabetes: A Predictor of Cardiometabolic Events. *Front. Neurosci.* **2018**, *12*, 591. [CrossRef]
65. Agashe, S.; Petak, S. Cardiac Autonomic Neuropathy in Diabetes Mellitus. *Methodist DeBakey Cardiovasc. J.* **2018**, *14*, 251. [CrossRef]
66. Benichou, T.; Pereira, B.; Mermillod, M.; Tauveron, I.; Pfabigan, D.; Maqdasy, S.; Dutheil, F. Heart Rate Variability in Type 2 Diabetes Mellitus: A Systematic Review and Meta–Analysis. *PLoS ONE* **2018**, *13*, e0195166. [CrossRef] [PubMed]
67. Mohamed, A.A.; Alawna, M. Important Role of Relaxation Techniques in Immune Functions, Glycemic Control, and Stress in Diabetic Patients with COVID-19: A Review. *Curr. Diabetes Rev.* **2021**, *17*, e121020186816. [CrossRef] [PubMed]
68. Vanzella, L.M.; Bernardo, A.F.B.; de Carvalho, T.D.; Vanderlei, F.M.; da Silva, A.K.F.; Vanderlei, L.C.M. Complexity of Autonomic Nervous System Function in Individuals with COPD. *J. Bras. Pneumol.* **2018**, *44*, 24–30. [CrossRef] [PubMed]
69. Hamasaki, H. Effects of Diaphragmatic Breathing on Health: A Narrative Review. *Medicines* **2020**, *7*, 65. [CrossRef]
70. Lachowska, K.; Bellwon, J.; Moryś, J.; Gruchała, M.; Hering, D. Slow Breathing Improves Cardiovascular Reactivity to Mental Stress and Health-Related Quality of Life in Heart Failure Patients with Reduced Ejection Fraction. *Cardiol. J.* **2020**, *27*, 772–779. [CrossRef]

71. Lachowska, K.; Bellwon, J.; Narkiewicz, K.; Gruchała, M.; Hering, D. Long-Term Effects of Device-Guided Slow Breathing in Stable Heart Failure Patients with Reduced Ejection Fraction. *Clin. Res. Cardiol.* **2019**, *108*, 48–60. [CrossRef]
72. Larson, M.; Chantigian, D.P.; Asirvatham-Jeyaraj, N.; Van de Winckel, A.; Keller-Ross, M.L. Slow-Paced Breathing and Autonomic Function in People Post-Stroke. *Front. Physiol.* **2020**, *11*, 573325. [CrossRef]
73. Luna-Rodríguez, G.L.; Peláez-Hernández, V.; Orea-Tejeda, A.; Ledesma-Ruíz, C.D.; Casarín-López, F.; Rosas-Trujillo, A.; Domínguez-Trejo, B.; Tepepa-Flores, L.E. Prevalence of Post-Traumatic Stress Disorder, Emotional Impairments, and Fear in COVID-19 Surviving Patients. *Front. Virtual Real.* **2022**, *3*, 927058. [CrossRef]
74. Egede, L.E.; Walker, R.J.; Dawson, A.Z.; Zosel, A.; Bhandari, S.; Nagavally, S.; Martin, I.; Frank, M. Short-Term Impact of COVID-19 on Quality of Life, Perceived Stress, and Serious Psychological Distress in an Adult Population in the Midwest United States. *Qual. Life Res.* **2022**, *31*, 2387–2396. [CrossRef]
75. Nalbandian, A.; Sehgal, K.; Gupta, A.; Madhavan, M.V.; McGroder, C.; Stevens, J.S.; Cook, J.R.; Nordvig, A.S.; Shalev, D.; Sehrawat, T.S.; et al. Post-Acute COVID-19 Syndrome. *Nat. Med.* **2021**, *27*, 601–615. [CrossRef]
76. Allendes, F.J.; Díaz, H.S.; Ortiz, F.C.; Marcus, N.J.; Quintanilla, R.; Inestrosa, N.C.; Del Rio, R. Cardiovascular and Autonomic Dysfunction in Long-COVID Syndrome and the Potential Role of Non-Invasive Therapeutic Strategies on Cardiovascular Outcomes. *Front. Med.* **2023**, *9*, 1095249. [CrossRef] [PubMed]
77. Fournié, C.; Chouchou, F.; Dalleau, G.; Caderby, T.; Cabrera, Q.; Verkindt, C. Heart Rate Variability Biofeedback in Chronic Disease Management: A Systematic Review. *Complement. Ther. Med.* **2021**, *60*, 102750. [CrossRef]
78. Lehrer, P.M.; Gevirtz, R. Heart Rate Variability Biofeedback: How and Why Does It Work? *Front. Psychol.* **2014**, *5*, 756. [CrossRef] [PubMed]
79. Gevirtz, R. The Promise of Heart Rate Variability Biofeedback: Evidence-Based Applications. *Biofeedback* **2013**, *41*, 110–120. [CrossRef]
80. Scheer, F.A.J.L.; Chellappa, S.L.; Hu, K.; Shea, S.A. Impact of Mental Stress, the Circadian System and Their Interaction on Human Cardiovascular Function. *Psychoneuroendocrinology* **2019**, *103*, 125–129. [CrossRef] [PubMed]
81. Shaffer, F.; McCraty, R.; Zerr, C.L. A Healthy Heart Is Not a Metronome: An Integrative Review of the Heart's Anatomy and Heart Rate Variability. *Front. Psychol.* **2014**, *5*, 1040. [CrossRef] [PubMed]
82. Goessl, V.C.; Curtiss, J.E.; Hofmann, S.G. The Effect of Heart Rate Variability Biofeedback Training on Stress and Anxiety: A Meta-Analysis. *Psychol. Med.* **2017**, *47*, 2578–2586. [CrossRef]

entropy

MDPI

Article

Pathological Heart Rate Regulation in Apparently Healthy Individuals

Ludmila Sidorenko [1], Irina Sidorenko [2], Andrej Gapelyuk [3] and Niels Wessel [3,4,*]

[1] Department of Molecular Biology and Human Genetics, State University of Medicine and Pharmacy, "Nicolae Testemitanu", Stefan cel Mare Str. 165, MD-2004 Chisinau, Moldova; ludmila.sidorenco@usmf.md
[2] Medical Center "Gesundheit", Mihai Kogalniceanu Str. 45/2, MD-2009 Chisinau, Moldova
[3] Cardiovascular Physics, Humboldt-Universität zu Berlin, D-10099 Berlin, Germany
[4] MSB Medical School Berlin GmbH, D-14197 Berlin, Germany
* Correspondence: wessel@physik.hu-berlin.de

Abstract: Cardiovascular diseases are the leading cause of morbidity and mortality in adults worldwide. There is one common pathophysiological aspect present in all cardiovascular diseases—dysfunctional heart rhythm regulation. Taking this aspect into consideration for cardiovascular risk predictions opens important research perspectives, allowing for the development of preventive treatment techniques. The aim of this study was to find out whether certain pathologically appearing signs in the heart rate variability (HRV) of an apparently healthy person, even with high HRV, can be defined as biomarkers for a disturbed cardiac regulation and whether this can be treated preventively by a drug-free method. This multi-phase study included 218 healthy subjects of either sex, who consecutively visited the physician at Gesundheit clinic because of arterial hypertension, depression, headache, psycho-emotional stress, extreme weakness, disturbed night sleep, heart palpitations, or chest pain. In study phase A, baseline measurement to identify individuals with cardiovascular risks was done. Therefore, standard HRV, as well as the new cardiorhythmogram (CRG) method, were applied to all subjects. The new CRG analysis used here is based on the recently introduced LF drops and HF counter-regulation. Regarding the mechanisms of why these appear in a steady-state cardiorhythmmogram, they represent non-linear event-based dynamical HRV biomarkers. The next phase of the study, phase B, tested whether the pathologically appearing signs identified via CRG in phase A could be clinically influenced by drug-free treatment. In order to validate the new CRG method, it was supported by non-linear HRV analysis in both phase A and in phase B. Out of 218 subjects, the pathologically appearing signs could be detected in 130 cases (60%), $p < 0.01$, by the new CRG method, and by the standard HRV analysis in 40 cases (18%), $p < 0.05$. Thus, the CRG method was able to detect 42% more cases with pathologically appearing cardiac regulation. In addition, the comparative CRG analysis before and after treatment showed that the pathologically appearing signs could be clinically influenced without the use of medication. After treatment, the risk group decreased eight-fold—from 130 people to 16 ($p < 0.01$). Therefore, progression of the detected pathological signs to structural cardiac pathology or arrhythmia could be prevented in most of the cases. However, in the remaining risk group of 16 apparently healthy subjects, 8 people died due to all-cause mortality. In contrast, no other subject in this study has died so far. The non-linear parameter which is able to quantify the changes in CRGs before versus after treatment is FWRENYI4 (symbolic dynamic feature); it decreased from 2.85 to 2.53 ($p < 0.001$). In summary, signs of pathological cardiac regulation can be identified by the CRG analysis of apparently healthy subjects in the early stages of development of cardiac pathology. Thus, our method offers a sensitive biomarker for cardiovascular risks. The latter can be influenced by non-drug treatments (acupuncture) to stop the progression into structural cardiac pathologies or arrhythmias in most but not all of the patients. Therefore, this could be a real and easy-to-use supplemental method, contributing to primary prevention in cardiology.

Keywords: heart rate variability; cardiorhythmogram; pathological appearing heart rhythm regulation; non-linear dynamics; risk prediction

1. Introduction

Cardiovascular diseases are the leading cause of morbidity and mortality in adults worldwide [1], although ongoing development is enabling more and more modern treatment techniques. Therefore, more resources should be invested in the prediction of structural heart diseases and arrhythmias, as well as in the research of prophylactic measurements [1–3]. A possible solution for prediction is offered by physiology. From a physiological point of view, before a structural heart disease or arrhythmia is manifested, these are anticipated by a latent period of disturbances of the heart regulation [4–6]. Usually, a structural heart disease is manifested when all of the compensatory mechanisms of regulation of the heart at the medullary and the central levels are broken down [7,8]. The person then enters a dangerous state, where the heart regulation migrates in a compensatory sense from the medullary to the predominant central level of regulation [4,9,10]. It is obvious that such a migration of heart regulation should be recognized as early as possible in order to prevent further progression into a structural heart disease or arrhythmia [3,8,11,12].

One of the well-known methods nowadays for assessing the state of the neural regulatory systems of the heart is the heart rate variability (HRV) [6,10,13–15]. Applying this method, when judging the reliability of the transition from the medullary level of regulation to the central one, shows a lot of limitations [8,13,16,17]. One of the important limitations is the appearance of non-steady-state events in a steady-state cardiorhytmogram (CRG—time series of beat-to beat-intervals; also called tachogram) [18,19]. This is because nonstationary HRV, analyzed when using standard linear methods, does not always show valid and credible results [8,13,16,18,19]. In the present study, a recently introduced CRG approach is offered by using dynamical physiological methods. Another essential problem when applying classical linear methods of HRV for the appreciation of the physiological state of the heart's regulation, e.g., for prognosis construction, is that it is mainly based on the conclusion: whether the HRV is high or low [16]. In this context, it is important to keep in mind that the HRV decreases significantly when all of the physiological compensatory mechanisms are broken down [6,11,12,20]. From a clinical point of view, this means that, under such conditions, the patient starts showing corresponding morbidity manifestations [11,21–23]. This is why a classical linear analysis cannot be applied for primary prophylaxis or a reliable prediction of cardiovascular risks in healthy individuals [3,12,24].

In this study, we searched for methods which would enable a reliable detection of biomarkers for pathological heart regulation in still-healthy persons even with a high HRV. Additionally, to support the realization of this aim, the CRGs were analyzed by non-linear methods. Non-linear methods [3,15,17,25–28] are recently introduced methods to measure HRV and are not affected by non-stationarity, as linear indexes are. They include a.o. power law exponent, approximate entropy, and detrended fluctuation analysis. These methods study all complex interactions of hemodynamic, electrophysiological, and humoral variables as well as the autonomic and central nervous regulations.

Therefore, primarily, the aim of this study was to find out whether certain signs which show a pathological cardiac regulation, detected in a CRG of an apparently healthy person, even with high HRV, can be influenced clinically. Secondly, the study aimed to find out which non-linear parameter can characterize the pathophysiological signs identified in CRG, supporting the automatic CRG assessment for risk prediction.

2. Material and Methods

In this study, 218 healthy persons were included, who consecutively visited the physician at medical center "Gesundheit" because of arterial hypertension, depression, headache, psycho-emotional stress, feeling of extreme weakness, disturbed night sleep, feeling of heart palpitations, or chest pain. However, physical examinations, i.e., ECG, stress test, and echocardiography, did not reveal any structural cardiovascular diseases. The general duration of the study was three years. Both genders were included, female 65% (N = 142). The average age was 42.8 ± 12.0 years, the BMI was 25.11 ± 4.9.

The inclusion criteria were echocardiographic healthy persons, i.e., persons without structural heart diseases in sinus rhythm during the measurement. The exclusion criteria were: age under 18 years, arrhythmias, structural heart diseases, pregnancy, lactation period, menopause, influenza, thyrotoxicosis, medication which could influence the HRV, Parkinson disease, and any kind of state where tremor of extremities was observed.

From all subjects included, the HRV was analyzed by linear as well as non-linear methods, including the pathophysiologic CRG features. Therefore, the biosignal recording was done in the form of a 5-min steady-state ECG. Hence, the ECG was recorded using the specialized hardware Polyspectrum. All standard rules and procedures during the measurement were applied [8,13,16]. During the preparation of the persons for the measurement, all standard rules and procedures were also respected, most prominently the reaching of steady state before starting the measurement itself [13,16,18]. The physiologic method of cardiorhythmogram analysis is based on features' analysis and is dependent on the person who analyzes the features. In order to approve this process and to create an automatized basis for the prediction of cardiovascular risks by analyzing a CRG, it was also analyzed by standard linear and existing non-linear methods of HRV analysis. The parameters applied for the linear methods of HRV analysis were meanNN, sdNN, sdaNN1, rmssd, pNN50, cvNN, VLF, LF, and HF [13,16,18,19]. To the non-linear HRV parameters belong Shannon, noNNtime, FORBWORD, FWSHANNON, FWRENYI4, WSDVAR, WPSUM02, and WPSUM13 [17,27–29]. FORBWORD stands for the symbolic dynamics parameter number of "forbidden words" in the distribution of words with length three: that is, the number of words which never or only seldomly occur. A high number of forbidden words stands for a rather regular behavior in the time series. If the time series is highly complex then, when using Shannon, only a few forbidden words will be found. FWSHANNON and FWRENYI4 denote the Shannon resp. Renyi entropy (of order 4) is calculated from the distribution of words in the same symbolic dynamics and both are classical complexity measures in time series. Higher entropy values refer to higher complexity in the corresponding tachograms and lower values to lower ones. WSDVAR is defined as the variability of different types of words occurring in the given symbolic dynamics. WPSUM02 and WPSUM13 quantify the percentage of words with low variability (02) as well as with high variability (13). The physiologic CRG analysis takes certain features of the waves of the heart rate time series—the low frequency (LF) drops and the high frequency (HF) exhibits counter-regulation. They can be quantified statistically by FWRENYI4 because the Renyi entropy of order 4 estimates the global complexity of the tachograms—rarely occurring words have no weight.

The study was performed in several stages. In stage A, all individuals were subject to the initial HRV measurement, followed by HRV analysis. This was performed step by step: firstly, the HRV was analyzed using standard linear methods; secondly, a physiological CRG analysis and non-linear analysis was performed. The aim of stage A was to divide all subjects into a group of healthy individuals and a risk group. The division was done based on the standard HRV as well as on the physiologic CRG method. According to the described HRV application standards [8,16], based on the linear HRV, the individuals were divided into the risk group if the HRV was low, and into the healthy group if the HRV was high [8,16,18]. HRV was defined as high if pNN50 > 0.14; it was considered as low if pNN50 $\leq$ 0.14. The physiologic CRG analysis allocated the persons with identified pathological signs in the cardiorhythmogram to the risk group. In the case of no pathological signs in the cardiorhythmogram, the individuals were classified into the healthy group.

In the next stage, stage B, it was proved whether the pathological signs identified in stage A can be influenced clinically by a drug-free treatment. It was important to find out whether the signs which were recognized as abnormal in the cardiorhythmograms, indicating a pathological state of the heart's regulation, could be treated before they would progress into structural heart disease or arrhythmia. Acupuncture was chosen as a physiological drug-free treatment. Therefore, all subjects from the risk group were treated.

In the next stage, stage C, a comparative measurement of HRV by linear, non-linear, and physiological CRG analysis, immediately following the end of the acupuncture treatment course, was done. The chosen treatment method corresponded to some important criteria: it should be a non-pharmacological treatment, a non-psychological and non-psychiatric treatment, and a treatment which corresponds to the demand of reproducibility and, henceforth, could be further applied in order to prevent the progression of the pathological signs. One type of possible treatment, corresponding to these criteria, is acupuncture [30,31]. In the study conducted by Sroka K., it is reported that acupuncture can be applied as a reliable drug-free alternative for restoring the physiological functional state of the parasympathical component of the VNS, being reflected in the stabilization of the heart rhythm [30]. Hence, all subjects from the risk group were treated with acupuncture. Acupuncture treatment in our study included 12–15 sessions. It is important to mention that we chose a non-medicamentous way of treatment because, in the risk group, healthy subjects were included who were still showing no structural cardiac morbidity manifestations or arrhythmia, but in whose cardiorhythmograms pathological signs were identified. Consequently, there was no evidence to apply any pharmaceutical way of treatment.

Then followed stage C: in order to objectively appreciate the changes in the pathological signs under the influence of treatment, each person from the pathological group was subject to an HRV investigation immediately after the acupuncture course. Then this was compared with the initial investigation, the baseline measurement, done before the acupuncture treatment. Therefore, all investigations after the treatment course were analyzed by the standard linear and non-linear HRV analysis, and by the new physiological CRG method. The comparative analysis of HRV and the cardiorhythmogram of the measurements done before and after the treatment should answer the following questions: Do the non-steady-state events actually disappear after the treatment? How is the non-linear parameter of the CRG method, the counter-regulation, changing after the course? How does the HRV change after the course?

At the concluding stage, stage D, a comparative analysis between both methods of analysis, in order to generate physiological biomarkers, was made based on the identified pathological signs during cardiorhythmogram analysis, supported by the non-linear analysis.

3. Description of the Study Design

Stage A: Initial HRV measurement

I Standard linear HRV analysis:

- Individuals with low HRV—the risk group;
- Individuals with high HRV—the group of healthy;
- NA individuals—those who could not be analyzed by these methods.

II Physiological cardiorhythmogram analysis:

- Individuals with identified pathological signs—the risk group;
- Individuals without identified pathological signs—the group of healthy;

Stage B: To prove whether the pathological signs identified in stage A can be influenced clinically by a drug-free treatment

Acupuncture—a physiological drug-free treatment (included 12–15 sessions).

All subjects from the risk group were treated with acupuncture.

Stage C: Comparative measurement of HRV at the end of the acupuncture treatment course.

Stage D: Comparative analysis between standard linear HRV analyzing method and physiological method of cardiorhythmogram analysis in order to generate physiological biomarkers based on the identified pathological signs, supported by non-linear analysis.

Statistical analysis was performed by Fischer's Exact Test. $p < 0.05$ was considered as significant, $p < 0.01$ means a very high significance. $p > 0.05$ was considered as not significant.

The present study was approved by the Research Ethics Committee at the session hold on 13 June 2016, the number of the document "Favorable Opinion of the Research Ethics Committee" is 78, dated 17 June 2016. The Chairperson of the Research Ethics Committee during the session was Prof. Viorel Nacu. The session took place at the "Nicolae Testemitanu" State University of Medicine and Pharmacy, Chisinau, Republic of Moldova.

4. Results

During the first stage, stage A, after the baseline measurement, all subjects (N = 218) were classified into the risk group and healthy group, based on two approaches to CRG analysis—the linear HRV method and the recent non-linear physiologic CRG method. In Figure 1, the comparative analysis of both methods which were used in this study in order to identify the pathological signs in cardiorhythmograms of healthy individuals is shown. Based on standard linear HRV analysis, 18% (N = 40) of individuals from the total amount N = 218 could be categorized into the risk group, whereas, 51% (N = 110) of individuals were categorized in the healthy group. In 68 individuals, the CRG could not be analyzed by the standard linear HRV method, so 32% (N = 68) of individuals were categorized into the non-available (NA) group. Based on the recent non-linear physiologic CRG method, 60% (N = 130) of individuals out of 218 were classified into the risk group, whereas 40% (N = 88) out of 218 were classified into the healthy group (Figure 1). By this approach, all CRGs could be analyzed, so that no one was classified into the non-available (NA) group.

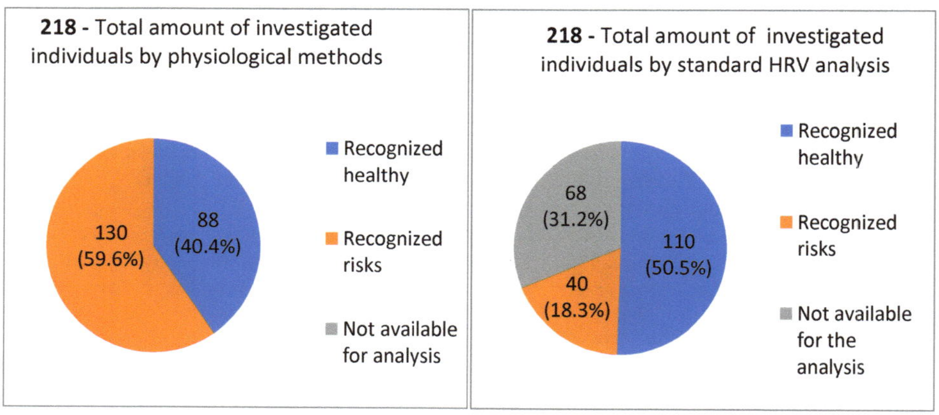

Figure 1. Results of baseline cardiorhythmograms analysis by the physiological method (**left**) and by the standard linear HRV method (**right**). Based on this analysis, the individuals were divided in two groups: the healthy group and the group with recognized risks.

Also, in the group of subjects who were classified by the linear methods initially as healthy, pathological signs were identified by the new physiological CRG method. Therefore, using the new method could also find those who were under the sensitivity threshold of the linear method. This is why the number of healthy individuals decreased from 110 to 88 ($p < 0.01$) after the physiological analysis. Correspondingly, the number of individuals with recognized risks increased from 40 to 133 ($p < 0.01$), as analyzed by the CRG method (Figure 1). In Figure 2 a schematical representation of the results obtained by applying the standard linear HRV method and the new physiological method of cardiorhythmogram analysis at every stage of the study: baseline measurement, after the treatment, and comparative analysis between both methods of analysis is shown (Figure 2).

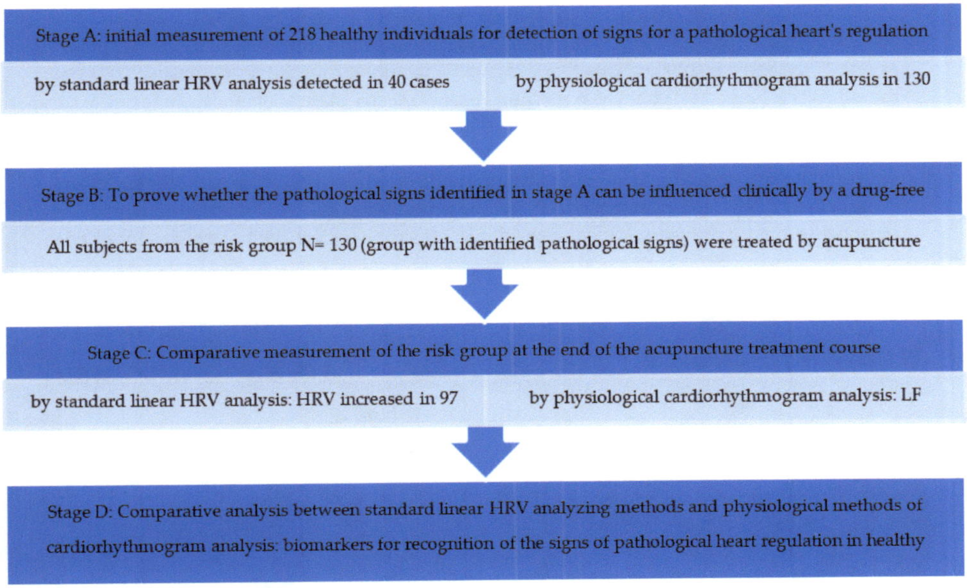

Figure 2. Scheme of results obtained by applying the standard linear HRV method and the new physiological method of cardiorhythmogram analysis at every stage of the study: baseline measurement, after the treatment, and comparative analysis between both methods of analysis.

Description of the scheme.
Stage A: baseline HRV measurement of 218 individuals
I Standard linear HRV analysis of 218

- Persons with low HRV—the risk group 40 (18%);
- Persons with high HRV—the group defined as healthy 110 (51%);
- NA Persons who could not be analyzed by this method 68 (32%).

II Physiological cardiorhythmogram analysis

- Persons with identified pathological signs—the risk group 130 (60%);
- Persons without identified pathological signs—the group defined as healthy 88 (40%).

Stage B: To prove whether pathological signs identified in stage A can be influenced clinically by a drug-free treatment with acupuncture. All subjects from the risk group (the group with identified pathological signs) were treated by acupuncture, so 130 from 218 were treated. The treatment course included 12–15 sessions.

Stage C: Comparative measurement of HRV at the end of the acupuncture treatment course.

Stage D: Comparative analysis between standard linear HRV analyzing methods and non-linear, including the recent physiological method of cardiorhythmogram analysis in order to generate physiological biomarkers based on the ability of identification of the pathological signs during HRV and the cardiorhythmogram analysis.

5. Clinical Influence on the Detected Signs of a Pathological Heart Regulation

The comparative analysis of HRV and the cardiorhythmogram of the measurements done before and after the treatment showed the following dynamic of non-steady-state events: in 114 ($p < 0.01$) of the cases, the LF drops were not visible in the cardiorhythmograms after the treatment course. This means that, from 130 subjects with the detected pathological signs, treated by acupuncture, in 114 (88%) $p < 0.01$ subjects any pathological signs were observed after the treatment course. Of 130 treated subjects, 12% (16) of the subjects still continued to show pathological signs (Figure 3). Another important parameter, which was evaluated by the physiological CRG analysis before versus after treatment, is the counter-regulation: the frequencies of counter-regulation changed from the LF waves to the HF waves in 81% (105) $p < 0.01$ of the cases. Such a change in counter-regulation is regarded as one of the most important positive prognostic factors. From the remaining 12% (16) of the subjects who had no changes regarding the non-steady-state events' dynamic of the LF drops and the counter-regulation, eight patients died. Representative cardiorhythmograms of two individuals, measured at baseline before and after the treatment (Figure 4). In Figure 4 the upper two CRGs belong to a patient in whose CRG the pathological signs disappeared after the treatment course and the other two CRGs (C, D) are from one of the 16 patients in who the pathological signs did not disappear after the treatment (deceased).

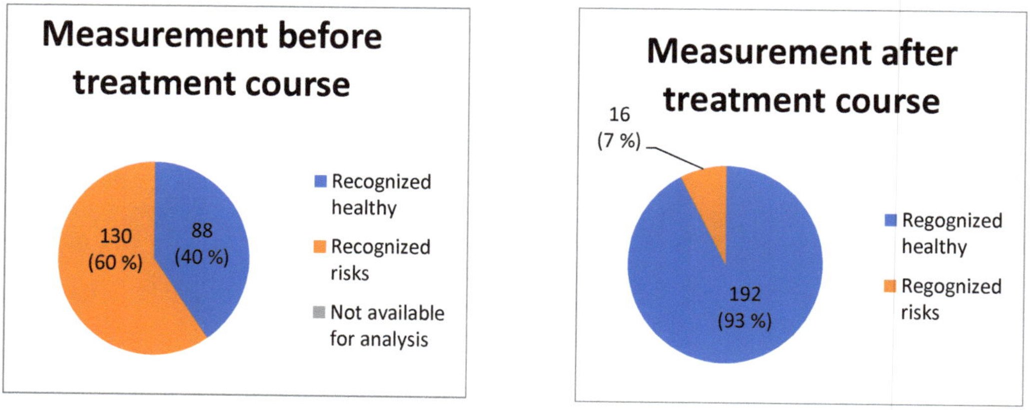

Figure 3. Results of comparative analysis between the CRGs before and after the treatment course, evaluating the presence of pathological signs. The effectiveness of the treatment course is clearly visible. The number of patients belonging to the risk group is reduced by eight times—from 130 individuals divided into the risk group before the treatment, the number decreased after the treatment to 16 ($p < 0.01$) individuals remaining in the risk group.

The physiological method of cardiorhythmogram analysis was objectively compared with standard linear and non-linear methods of HRV analysis. The non-linear method of HRV analysis more quantitatively described the changes in the cardiorhythmograms of patients before versus after treatment, in comparison to linear methods of HRV analysis. Among the linear methods which showed significant changes in the 114 patients, comparing cardiorhythmograms before and after treatment, were the following parameters (indicated as the mean values): meanNN changed from 890.15 to 947.43 ($p < 0.05$), sdNN changed from 40.23 to 47.82 ($p < 0.01$), sdaNN1changed from 15.16 to 24.03 ($p < 0.01$), rmssd changed from 32.28 to 38.69 ($p < 0.01$), and pNN50 changed from 0.13 to 0.19 ($p < 0.01$). In conclusion, the changes of the HRV by linear methods, comparing the measurement before and after the treatment course, were: the HRV increased significantly after the treatment in 75% (N = 97) $p < 0.01$ of cases. In 24% (N = 31) $p < 0.05$ of cases, the HRV remained without any changes. In 1.5% (N = 2) of cases the HRV decreased.

Among the non-linear methods which showed significant changes in the 114 patients, comparing cardiorhythmograms before and after treatment, were the following parameters (indicated as the mean values): Shannon changed from 1.95 to 2.15 ($p < 0.001$), FORBWORD from 13.96 to 19.48 ($p < 0.01$), FWSHANNON from 3.41 to 3.19 ($p < 0.001$), FWRENYI4 from 2.85 to 2.53 ($p < 0.001$), WSDVAR from 2.33 to 2.48 ($p < 0.001$), WPSUM02 from 0.02 to 0.01 ($p < 0.01$), and WPSUM13 from 0.41 to 0.51 ($p < 0.001$).

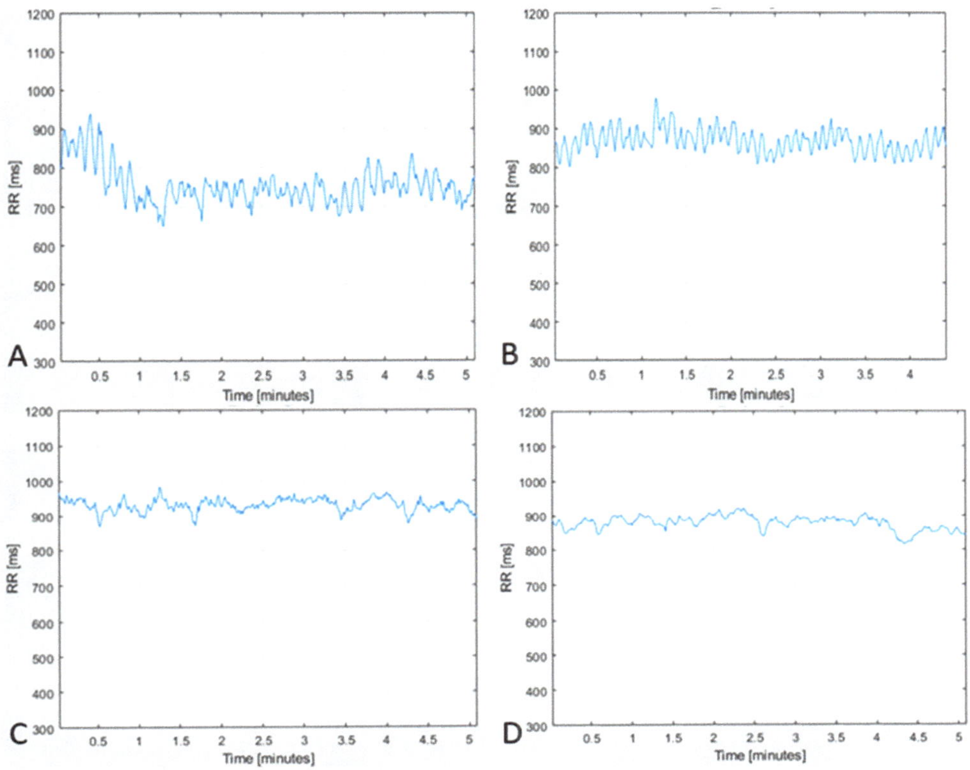

Figure 4. Representative cardiorhythmograms of two individuals, measured at baseline before and after the treatment. The CRGs (**A**,**B**) belong to a patient in whose CRG the pathological signs disappeared after the treatment course. The CRGs (**C**,**D**) are from one of the 16 patients in who the pathological signs did not disappear after the treatment (deceased). Both (**A**,**C**) show pathological signs in the form of LF drops, followed by a pathological counter-regulation. After the treatment course in (**B**), the pathological signs disappeared. In (**D**), the pathological signs still remain.

As can be seen from the comparative analysis of the cardiorhythmograms before and after the acupuncture course, it is evident that the pathological signs in the cardiorhythmograms found in this study can be detected by physiological CRG analysis. Secondly, pathological signs can be influenced clinically (Figure 4), without using drugs. Following the treatment with acupuncture, a large number (88%) N = 114 ($p < 0.01$) of subjects from the risk group could move to the healthy group after the treatment course. Like the results show (Figure 3), a progression of these pathological signs into structural heart pathology or arrhythmia could be prevented in the majority of cases.

6. Discussion

During the last 30 years, the morbidity and mortality of cardiovascular risks has drastically increased [1,23,32]. This fact has generated a challenging aim for researchers all over the world to find predictors for several cardiovascular risks. This scientific challenge can be solved interdisciplinarily, and the cooperation of physicians, mathematicians, physicists, and biostatistics is required. Since 2000, until now, a lot of attempts to propose predictors for several cardiovascular diseases have been made. Regarding the fact that, in every cardiovascular pathology, the regulatory cardiac system is involved, the approach of using non-linear dynamics in ECG and HRV analysis to find certain predictors has been used, e.g., it can be obviously observed, as shown by published studies on the prediction of atrial fibrillation and heart failure, by applying the non-linear method of analysis [15,17,23,25,26,32,33]. In these studies, there are different ECG features described as potential predictors. Zong et al. [25] showed, when observing a 30-min ECG, the feature which occurs before paroxysms of atrial fibrillation is the frequency of atrial premature contractions. In the study of Langley et al. [32], the possibility of predicting the recurrence of atrial fibrillation based on the assessment of a number of atrial and ventricular ectopic heartbeats is described. Therefore, they used 30-min RR interval data. A further prediction method regarding the appearance of recurrences of atrial fibrillation that is often described in the literature is the method of P-wave assessment on ECG [26,28,33]. In the studies which apply this method, researchers often take the duration of the P-wave into account, as well as its amplitude and the change in the P-wave, the intensities of the P-wave change power spectrum, and several non-linear P-wave measurements. In particular, Martinez A. et al. [33] describe the prediction of recurrence of atrial fibrillation, based on the P-waves, assessed using events observed one hour before the recurrence onset, as an effective tool. There are a lot of further examples. The findings of our study correspond to the literature, and the CRG features taken into consideration by the given study are also based on certain important pathophysiological mechanisms of heart regulation which stand at the background of the feature; these will be characterized further in this text. The advantages of the method proposed in this study consist in the possibility to recognize risks even in apparently healthy individuals, before the onset of disease or arrhythmia; therefore, they can be applied as predictors and, consequently, even measures of primary prophylaxis can be realized in the clinic. Another important advantage consists in the fact the CRG method can be assessed either by a trained physician, analyzing a CRG, or automatically, possessing a certain software, by evaluation of the CRG features via the non-linear parameter FWRENYI4.

The results of this study show that it is obvious that the physiological CRG method was more successful, in comparison with standard linear HRV analysis, in detecting pathological signs in cardiorhythmograms of apparently healthy individuals. Comparing the physiological method of cardiorhythmogram analysis with the standard linear HRV analysis, the recognition of pathological signs in the cardiorhythmograms of healthy individuals was about 60%, N = 130 from 218 cases ($p < 0.01$), compared to 18%, N = 40 from 218 cases ($p < 0.05$) recognized by the standard linear HRV analysis. This means that the sensitivity of the new method is three times higher in comparison with the existing method of standard linear HRV analysis. One of the reasons that explains such a remarkable difference is the limitation of the standard linear HRV analysis [16,18,19], which does not allow the standard HRV to analyze CRGs with events of non-stationarity, despite the fact that events of non-stationarity occurring in a steady-state CRG reveal predictively important information regarding the cardiac regulation [4,27,34,35]. The non-linear methods, including the new CRG method, do have the possibility to analyze such CRGs [15,25,27–29,32]. The limitations also belong to the factors, which explains why the number of the NA group was high in the case of linear HRV analysis. The NA group includes all of the ECGs which could not be analyzed by linear methods. This amounted to about 32%. This means that, in 68 out of 218 individuals, the CRGs could not be analyzed by using the linear HRV analysis. Taking into account that pNN50 can be, anyhow, assessed [13,16,18], it should be mentioned that, for a high number of the non-stationarity events, the LF drops and extra beats in the CRGs of the individuals of the NA group, a valid interpretation of pNN50 was also not possible [16,18,19]. In comparison, there was not one individual who could not be analyzed by using the physiological method, so that all 218 subjects of the study could be analyzed by using this method. To conclude, the new method, the physiological method of cardiorhythmogram analysis, allowed for the analysis even of those subjects who could not be assessed by the standard linear HRV method. The fact that, due to the CRG method, all cardiorhythmograms can be analyzed, shows the high reliability of the method. The number of individuals who were considered healthy by the linear method decreased after the analysis by the physiological method, from 110 to 88 individuals. This means that, by using the physiological method of CRG analysis, pathological signs can be recognized even in cases where the linear method does not detect these.

Validity is another very important characteristic of the recent CRG method, from a clinical point of view. It is manifested by the assessment results of the dynamics of pathological signs—the LF drops and HF counter-regulation before and after the treatment. Recently, it is one of the few methods which can assess the dynamic changes in the quality of waves' frequencies in HF counter-regulation and LF drops [4,24,27]. It is clearly visible how precisely the new HF counter-regulation parameter evaluated the CRG before versus after the treatment. It could be recognized that the frequencies of counter-regulation changed from the LF waves to the HF waves in 81% (105) $p < 0.01$ of the cases. Such a change in the counter-regulation is regarded as one of the most important positive prognostic factors. Therefore, the physiological CRG method is a sensitive assessment tool which is able to deliver a valid evaluation of the changes in pathological signs even in healthy individuals. Furthermore, the pathological signs can serve as biomarkers for the test whether the risk disappeared or not, i.e., whether the treatment for this person was efficient or not—is the person out of the risk group or does he still belong to the risk group? LF drops is the second parameter of the CRG method. It makes risk stratification possible via its presence in a steady-state CRG. The assessment of its dynamic offers the possibility to understand whether the person is still in the risk group. The results of the present study show that, in 114 ($p < 0.01$) cases, the LF drops were absent in the cardiorhythmograms after the treatment course. This means that, after the treatment course for 130 subjects with the detected pathological signs, in 114 (88%) $p < 0.01$ individuals, any pathological signs were observed. Thus, the pathological signs can be influenced clinically by non–drug treatment. After the treatment, 93% of the individuals from the risk group were transferred to the healthy group. Therefore, the recognized cardiovascular risks in healthy individuals can be eliminated, preventing progression into structural heart disease or arrhythmia. Of 130 treated subjects, 12% (16) of the subjects still continued to show pathological signs (Figures 3 and 4).

A very important proof of the validity of the new CRG method is the fact that, from the remaining 12% (16) of the subjects who had no changes regarding the non-stationary events' dynamic of the LF drops and the counter-regulation, eight patients died. This also means that the pathological signs—the LF drops and HF counter-regulation—can be clinically applied as biomarkers for the identification of individuals who are at high risk. The important applicative value of the CRG method is represented on the one hand by risk prediction and on the other hand by the possibility of risk prevention. A further study with a higher number of individuals is needed to prove the predictive values of the pathological signs. The changes in the CRGs, which could be assessed by the recent CRG method, were also objectively proven by the non-linear parameter FWRENYI4. It validated the results of CRG method with high significance methodological differences between the standard linear method of HRV analysis and the physiological CRG method for the assessment of heart regulation.

For the linear method of analysis, a steady-state cardiorhythmogram (Figure 5) was analyzed by time-domain methods [13,16,18]. When in a cardiorhythmogram, non-steady-state events (Figure 6) occurred, such as the cardiorhythmogram having to be excluded from the analysis by the linear methods [16,18,19]. It is important to know that, especially in cardiorhythmograms, where non-steady-state events in rest state occur, important information is hidden regarding the detection of pathological signs in healthy individuals [4,10,15,19,22–24]. Such non-steady-state events in a rest-state cardiorhythmogram are described in the research work of N. Wessel [27]. In his work, these events are called "grosse Ereignisse" (dt.—big events). He found them in the resting state of patients who had an increased risk for developing a heart attack. In order to analyze them, he used non-linear methods of HRV analysis. In another study, the non-steady-state events in the cardiorhythmograms are called LF drops [4,24] and are described as risk factors for atrial fibrillation. In our study, these events in cardiorhythmograms were analyzed by the physiological CRG method [4] in order to find out whether these can be biomarkers for cardiovascular risks in healthy individuals.

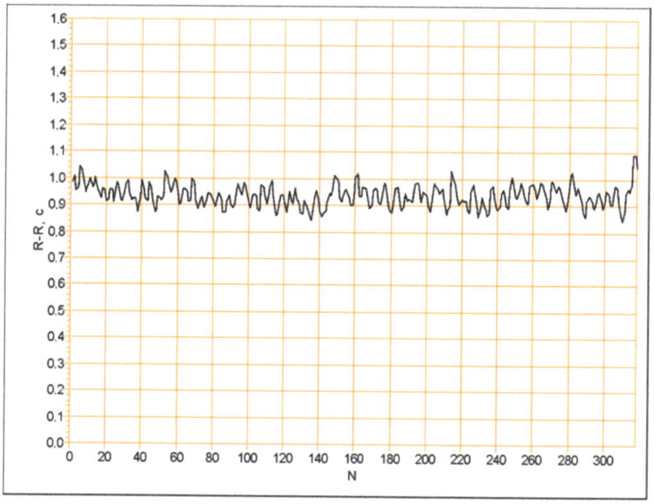

Figure 5. Steady-state cardiorhythmogram without events of non-stationarity, suitable for standard linear HRV analysis.

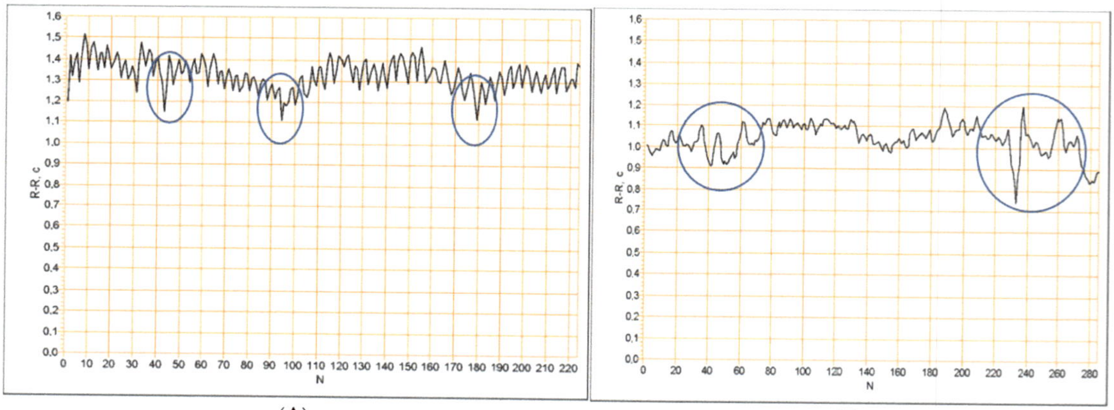

(**A**) (**B**)

Figure 6. Steady-state cardiorhythmograms where events of non-stationarity are presented. Cardiorhythmogram (**A**): the pathological signs are presented by the LF drops (encircled), but they are followed by a physiological counterbalancing via HF waves, so a low risk is estimated. Cardiorhythmogram (**B**): LF drops (encircled) are present followed by a pathological counter-regulation, predominant by LF waves, high risk is estimated.

The new physiological cardiorhythmogram analysis used here is based on the recently introduced LF drops and HF counter-regulation, which are non-linear, event-based dynamical HRV biomarkers [24]. The neural regulation of the heart rhythm is the physiological background of this method [4–6,14,20]. It is visualized by the corresponding waves' structure in a cardiorhythmogram [4,24]. When analyzing a cardiorhythmogram by using physiological methods, there are some essential rules to be regarded. First, the cardiorhythmogram should be checked for whether it contains non-steady-state events (Figure 6), as the appearance of these in a rest-state cardiorhythmogram is characterized as a pathological sign [6,23,24,36]. From a pathophysiological point of view, this means that, in the rest state, the central neuronal heart regulation dominates the heart modulation versus the medullary regulation [4,9,10,20,37,38]. Using another way, the central neural heart regulation is increased to being pathologically high and the activity of the medullary heart regulation is insufficient [4,9,10,15,20,37,38]. Such a state should be recognized as early as possible because its progression can lead to structural heart diseases and arrhythmias [11,14,21,22,25,32].

In order to assess the level of progression of such a pathological state, the counter-regulation should be taken into account. The counter-regulation represents the part of the cardiorhythmogram following a non-stationarity event [4,24]. If the counter-regulation functions physiologically fine, it is driven by the parasympathetic part of the vegetative nervous system [4,20,21]. This is indicated by a certain wave structure (Figure 6, cardiorhythmogram A). In this case, in the cardiorhythmogram, high-frequency waves during the counter-regulation are recognized. Hence, the compensatory function of counter-regulation functions well. This means that it is able to compensate, during a certain period of time, the pathologically increased central neural heart regulation. The latter is presented by non-steady-state events in a cardiorhythmogram (Figure 6). If the counter-regulation was driven by the sympathetic part of the vegetative nervous system, this is considered as a pathological counter-regulation (Figure 6, cardiorhythmogram B) [4,21,34,35,37]. This means that the physiological compensation of the state of a pathologically increased central neural heart regulation cannot be compensated effectively enough [4,6,9,24,26,37] and is regarded as a risk factor for the appearance of arrhythmias, progression into structural heart diseases, and, in the worst-case scenario, death [6,11,14,22–25,27], as was also shown by the results of the present study.

In summary, the non-steady-state events identified in steady-state cardiorhythmograms of healthy subjects were regarded as pathological signs. These can be identified by physiological methods of HRV analysis. The pathological signs can be clinically influenced—after the treatment course, the pathological signs in the form of the non-steady-state events disappeared from the cardiorhythmogram, the HRV increased, and, in the majority of cases, the counter-regulation changed from sympathetically driven to parasympathetically driven. These changes led to the transfer of 88% (N = 114) of the subjects after the treatment from the pathological group to the healthy group. This has a very important clinical applicative value. Pathological signs can be identified in physiological cardiorhythmograms of healthy individuals in the early stages of the development of cardiologic pathology by applying non-complicated, easy, and quick methods of registration. Thus, they can be influenced by non-drug treatment methods in order to stop their progression into structural heart diseases or arrhythmia. Hence, this could be a real and easy application method for contributing to primary prophylaxis in cardiology.

The results show that it was possible to answer the question raised in this study in a positive way. Therefore, such pathological signs could further be defined as biomarkers for some cardiovascular risks and could, hence, serve for their prediction. This result's applicative value would be important for the primary prophylaxis of arrhythmias, some structural heart diseases, and also a potential heart attack [8,11,12,22–24,27]. Moreover, the predictive effectiveness of such biomarkers could be tested in a separate study on a specific cardiological disease. FWRENYI4 is the non-linear parameter that makes a significant evaluation of the changes in the pathophysiological signs identified in CRG possible. Further studies should research whether it can be applied for supporting the automatic CRG assessment for risk prediction.

7. Limitations of the Study

The limitation of this study is that the new CRG method, based on evaluation of the heart's regulation in the CRG and the recognition of signs for a pathologic heart regulation, depends on the skills of a trained physiologist. Although, in this study, the method was proven objectively and significantly by the non-linear parameter—the FWRENYI4, a further study to ensure the automaticity of the evaluation and prediction process is needed. The basis for such an automatized process is the non-linear parameter FWRENYI4. As was mentioned above, it approved, in this study, the evaluation technique by CRG method. Therefore, a further study should be done to test the predictive capacity of FWRENYI4 on a concrete clinical pathology.

8. Conclusions

1. In steady-state cardiorhythmograms of healthy individuals, pathological signs were identified that can be applied as biomarkers for the functional pathological state of the heart's regulation.
2. The efficacy of the new-found physiological method of cardiorhythmogram analysis for the recognition of cardiovascular risks in healthy individuals is significantly higher than the standard HRV analysis: using the new CRG method, the risk group detection is 60% versus 18% risk detection by the existing standard HRV linear analysis. Therefore, the sensitivity of the new method is three times higher in comparison with the existing method of standard linear HRV analysis.
3. The pathological signs on which the new physiological approach to the cardiorhythmogram analysis is based can be defined as biomarkers for cardiovascular risk recognition in healthy individuals.
4. The physiological CRG method is able to analyze even those cardiorhythmograms which show non-steady-state events.
5. The pathological signs can be influenced clinically by non–drug treatment. After the treatment, 93% of the individuals from the risk group were transferred to the healthy group. Therefore, the recognized cardiovascular risks in healthy individuals can be eliminated, preventing progression into structural heart disease or arrhythmia.
6. The non-linear parameter which significantly characterizes the dynamic of changes in the pathological signs in the CRGs before treatment versus after treatment is FWRENYI4.

Author Contributions: Conceptualization, I.S., L.S. and N.W.; methodology, L.S. and N.W.; software, L.S., N.W. and A.G.; validation, N.W. and A.G.; formal analysis, N.W., A.G. and L.S.; investigation, I.S., L.S.; patients' treatment I.S.; resources I.S. and N.W.; data curation, L.S.; writing—original draft preparation, L.S.; writing—N.W., A.G. and L.S.; visualization, N.W. and A.G.; supervision, N.W.; project administration, N.W. and L.S. All authors have read and agreed to the published version of the manuscript.

Funding: This research received no external funding.

Institutional Review Board Statement: The study was conducted in accordance with the Declaration of Helsinki, and approved by the Institutional Research Ethics Committee of the State University of Medicine and Pharmacy "Nicolae Testemitanu", Stefan cel Mare str. 165, MD-2004, Chisinau, Moldova (protocol code 78 and date of approval is 17 June 2016).

Data Availability Statement: Original data are available upon request from the first author.

Acknowledgments: In memoriam to our cardiological mentor Gert Baumann.

Conflicts of Interest: The authors declare no conflict of interest.

References

1. Goldsborough, E.; Ngozi, O.; Blaha, M.J. Assessment of Cardiovascular Disease Risk: A 2022 Update. *Endocrinol. Metab. Clin.* **2022**, *51*, 483–509. [CrossRef] [PubMed]
2. Matusik, P.S.; Matusik, P.T.; Stein, P.K. Heart rate variability and heart rate patterns measured from wearable and implanted devices in screening for atrial fibrillation: Potential clinical and population-wide applications. *Eur. Heart J.* **2023**, *44*, 1105–1107. [CrossRef] [PubMed]
3. Altini, M.; Plews, D. What Is behind Changes in Resting Heart Rate and Heart Rate Variability? A Large-Scale Analysis of Longitudinal Measurements Acquired in Free-Living. *Sensors* **2021**, *21*, 7932. [CrossRef] [PubMed]
4. Ludmila, S.; Ivan, D.R.; Victor, V.; Gert, B. Fundamental aspects of cardiovascular regulation in predisposition to atrial fibrillation. *Mold. Med. J.* **2018**, *61*, 42–45. [CrossRef]

5. Gitler, A.; Vanacker, L.; De Couck, M.; De Leeuw, I.; Gidron, Y. Neuromodulation Applied to Diseases: The Case of HRV Biofeedback. *J. Clin. Med.* **2022**, *11*, 5927. [CrossRef]
6. Grégoire, J.M.; Gilon, C.; Carlier, S.; Bersini, H. Autonomic nervous system assessment using heart rate variability. *Acta Cardiol.* **2023**, 1–15. [CrossRef]
7. Rudenko, M.; Zernov, V.; Voronova, O. Fundamental Research on the Mechanism of Cardiovascular System Hemodynamics Self-Regulation and Determination of the Norm-Pathology Boundary for the Basic Hemodynamic Parameters and Analysis of the Compensation Mechanism as a Method of Revealing the Underlying Causes of the Disease. *Heart Rhythm.* **2012**, *9*, 1909–1910. [CrossRef]
8. Catai, A.M.; Pastre, C.M.; de Godoy, M.F.; da Silva, E.; Takahashi, A.C.D.M.; Vanderlei, L.C.M. Heart rate variability: Are you using it properly? Standardisation checklist of procedures. *Braz. J. Phys. Ther.* **2020**, *24*, 91–102. [CrossRef]
9. Matusik, P.S.; Zhong, C.; Matusik, P.T.; Alomar, O.; Stein, P.K. Neuroimaging Studies of the Neural Correlates of Heart Rate Variability: A Systematic Review. *J. Clin. Med.* **2023**, *12*, 1016. [CrossRef]
10. Yang, L.; Wei, X.; Liu, F.; Zhu, X.; Zhou, F. Automatic feature learning model combining functional connectivity network and graph regularization for depression detection. *Biomed. Signal Process. Control.* **2023**, *82*, 104520. [CrossRef]
11. Florea, V.G.; Cohn, J.N. The Autonomic Nervous System and Heart Failure. *Circ. Res.* **2014**, *114*, 1815–1826. [CrossRef]
12. Souza, H.C.D.; Philbois, S.V.; Veiga, A.C.; Aguilar, B.A. Heart Rate Variability and Cardiovascular Fitness: What We Know so Far. *Vasc. Health Risk Manag.* **2021**, *17*, 701–711. [CrossRef]
13. Shaffer, F.; Ginsberg, J.P. An Overview of Heart Rate Variability Metrics and Norms. *Front. Public Health* **2017**, *5*, 258. [CrossRef]
14. Vovc, V.; Moldovanu, I.; Sidorenko, L.; Ganenco, A. Modificarea variabilităţii cardiace şi a paternului respirator prin stări psihoemoţional evocate. In *Anale Ştiinţifice ale USMF "Nicolae Testemiţanu"*; Ed. a 13-a.; CEP Medicina: Chişinău, Moldova, 2012; Volume 1, pp. 150–157.
15. Tiwari, R.; Kumar, R.; Malik, S.; Raj, T.; Kumar, P. Analysis of Heart Rate Variability and Implication of Different Factors on Heart Rate Variability. *Curr. Cardiol. Rev.* **2021**, *17*, e160721189770. [CrossRef]
16. Task Force of the European Society of Cardiology the North American Society of Pacing Electrophysiology. Heart Rate Variability: Standards of Measurement, Physiological Interpretation, and Clinical Use. *Circulation* **1996**, *93*, 1043–1065. [CrossRef]
17. Mueller, A.; Kraemer, J.F.; Penzel, T.; Bonnemeier, H.; Kurths, J.; Wessel, N. Causality in physiological signals. Topical Review. *Physiol. Meas.* **2016**, *37*, R46–R72. [CrossRef]
18. Billman, G.E.; Huikuri, H.V.; Sacha, J.; Trimmel, K. An introduction to heart rate variability: Methodological considerations and clinical applications. *Front. Physiol.* **2015**, *6*, 55. [CrossRef]
19. Sassi, R.; Cerutti, S.; Lombardi, F.; Malik, M.; Huikuri, H.V.; Peng, C.-K.; Schmidt, G.; Yamamoto, Y.; Gorenek, B.; Lip, G.Y.; et al. Advances in heart rate variability signal analysis: Joint position statement by the e-Cardiology ESC Working Group and the European Heart Rhythm Association co-endorsed by the Asia Pacific Heart Rhythm Society. *Europace* **2015**, *17*, 1341–1353. [CrossRef]
20. Andresen, M.C.; Mendelowitz, D. Autonomic Nervous System: Central Cardiovascular Control. In *Encyclopedia of Neuroscience*; Squire, L.R., Ed.; Academic Press: Cambridge, MA, USA, 2009; pp. 863–869, ISBN 9780080450469.
21. Esler, M. The sympathetic regulation of the heart. *Eur. Heart J.* **2016**, *37*, 2808–2809. [CrossRef]
22. Karoly, P.J.; Stirling, R.E.; Freestone, D.R.; Nurse, E.S.; Maturana, M.I.; Halliday, A.J.; Neal, A.; Gregg, N.M.; Brinkmann, B.H.; Richardson, M.P.; et al. Multiday cycles of heart rate are associated with seizure likelihood: An observational cohort study. *Ebiomedicine* **2021**, *72*, 103619. [CrossRef]
23. Mohebbi, M.; Ghassemian, H. Prediction of paroxysmal atrial fibrillation based on non-linear analysis and spectrum and bispectrum features of the heart rate variability signal. *Comput. Methods Programs Biomed.* **2012**, *105*, 40–49. [CrossRef] [PubMed]
24. Sidorenko, L.; Diaz-Ramirez, I.; Vovc, V.; Baumann, G. New approach to heart rate variability analysis based on cardiophysiological biomarkers. *Mold. Med. J.* **2018**, *61*, 39–46. [CrossRef]
25. Ros, E.; Mota, S.; Fernández, F.; Toro, F.; Bernier, J. ECG Characterization of paroxysmal atrial fibrillation: Parameter extraction and automatic diagnosis algorithm. *Comput. Biol. Med.* **2004**, *34*, 679–696. [CrossRef] [PubMed]
26. Boon, K.; Khalil-Hani, M.; Malarvili, M.; Sia, C. Paroxysmal atrial fibrillation prediction method with shorter HRV sequences. *Comput. Methods Programs Biomed.* **2016**, *134*, 187–196. [CrossRef]
27. Wessel, N.; Malberg, H.; Bauernschmitt, R.; Kurths, J. Nonlinear methods of cardiovascular physics and their clinical applicability. *Int. J. Bifurc. Chaos* **2007**, *17*, 3325–3371. [CrossRef]
28. Henriques, T.; Ribeiro, M.; Teixeira, A.; Castro, L.; Antunes, L.; Costa-Santos, C. Nonlinear Methods Most Applied to Heart-Rate Time Series: A Review. *Entropy* **2020**, *22*, 309. [CrossRef]
29. Işler, Y.; Kuntalp, M. Combining classical HRV indices with wavelet entropy measures improves to performance in diagnosing congestive heart failure. *Comput. Biol. Med.* **2007**, *37*, 1502–1510. [CrossRef]
30. Sroka, K. Der Parasympathikus in der Akupunktur und die Bedeutung seiner Aktivierung fuer die Herzgesundheit. German Journal of Acupuncture and Related Techniques. *Springer Med.* **2019**, *62*, 3–8.
31. Sidorenko, L.; Sidorenko, I. Novel physiological correction methods in functional disorders of heart rhythm. In Proceedings of the ISME 2019 Conference on Biomedical Engineering 2019, Haifa, Israel, 25–26 February 2019.
32. Narin, A.; Isler, Y.; Ozer, M.; Perc, M. Early prediction of paroxysmal atrial fibrillation based on short-term heart rate variability. *Phys. A Stat. Mech. Appl.* **2018**, *509*, 56–65. [CrossRef]

33. Martínez, A.; Abásolo, D.; Alcaraz, R.; Rieta, J.J. Alteration of the P-wave non-linear dynamics near the onset of paroxysmal atrial fibrillation. *Med. Eng. Phys.* **2015**, *37*, 692–697. [CrossRef]
34. Wessel, N.; Sidorenko, L.; Kraemer, J.; Schoebel, C.; Baumann, G. Assessing cardiac autonomic function via heart rate variability analysis requires monitoring respiration. *Europace* **2016**, *18*, 1280. [CrossRef]
35. Sidorenko, L.; Kraemer, J.K.; Wessel, N. Standard heart rate variability spectral analysis: Does it purely assess cardiac autonomic function? *Europace* **2016**, *18*, 1085. [CrossRef]
36. Castro, H.; Garcia-Racines, J.D.; Bernal-Norena, A. Methodology for the prediction of paroxysmal atrial fibrillation based on heart rate variability feature analysis. *Heliyon* **2021**, *7*, e08244. [CrossRef]
37. Pfurtscheller, G.; Blinowska, K.J.; Kaminski, M.; Rassler, B.; Klimesch, W. Processing of fMRI-related anxiety and information flow between brain and body revealed a preponderance of oscillations at 0.15/0.16 Hz. *Sci. Rep.* **2022**, *12*, 9117. [CrossRef]
38. McCraty, R.; Shaffer, F. Heart Rate Variability: New Perspectives on Physiological Mechanisms, Assessment of Self-regulatory Capacity, and Health risk. *Glob. Adv. Health Med.* **2015**, *4*, 46–61. [CrossRef]

entropy

MDPI

Article

Information-Theoretic Analysis of Cardio-Respiratory Interactions in Heart Failure Patients: Effects of Arrhythmias and Cardiac Resynchronization Therapy

Mirjana M. Platiša [1,*], Nikola N. Radovanović [2], Riccardo Pernice [3], Chiara Barà [3], Siniša U. Pavlović [2] and Luca Faes [3,*]

1 Laboratory for Biosignals, Institute of Biophysics, Faculty of Medicine, University of Belgrade, Višegradska 26-2, 11000 Belgrade, Serbia

2 Pacemaker Center, University Clinical Center of Serbia, University of Belgrade, 11000 Belgrade, Serbia; nikolar86@gmail.com (N.N.R.); pavlosini@yahoo.com (S.U.P.)

3 Department of Engineering, University of Palermo, Viale delle Scienze, Building 9, 90128 Palermo, Italy; riccardo.pernice@unipa.it (R.P.); chiara.bara@community.unipa.it (C.B.)

* Correspondence: mirjana.platisa@med.bg.ac.rs (M.M.P.); luca.faes@unipa.it (L.F.)

Abstract: The properties of cardio-respiratory coupling (CRC) are affected by various pathological conditions related to the cardiovascular and/or respiratory systems. In heart failure, one of the most common cardiac pathological conditions, the degree of CRC changes primarily depend on the type of heart-rhythm alterations. In this work, we investigated CRC in heart-failure patients, applying measures from information theory, i.e., Granger Causality (*GC*), Transfer Entropy (*TE*) and Cross Entropy (*CE*), to quantify the directed coupling and causality between cardiac (*RR interval*) and respiratory (*Resp*) time series. Patients were divided into three groups depending on their heart rhythm (sinus rhythm and presence of low/high number of ventricular extrasystoles) and were studied also after cardiac resynchronization therapy (CRT), distinguishing responders and non-responders to the therapy. The information-theoretic analysis of bidirectional cardio-respiratory interactions in HF patients revealed the strong effect of nonlinear components in the *RR* (high number of ventricular extrasystoles) and in the *Resp* time series (respiratory sinus arrhythmia) as well as in their causal interactions. We showed that *GC* as a linear model measure is not sensitive to both nonlinear components and only model free measures as *TE* and *CE* may quantify them. CRT responders mainly exhibit unchanged asymmetry in the *TE* values, with statistically significant dominance of the information flow from *Resp* to *RR* over the opposite flow from *RR* to *Resp*, before and after CRT. In non-responders this asymmetry was statistically significant only after CRT. Our results indicate that the success of CRT is related to corresponding information transfer between the cardiac and respiratory signal quantified at baseline measurements, which could contribute to a better selection of patients for this type of therapy.

Keywords: cardio-respiratory coupling; heart failure; ventricular extrasystoles; CRT responders; information theory; Granger Causality; Transfer Entropy; Cross Entropy

1. Introduction

The coordination between beat-to-beat (*RR*) interval and respiratory signals is crucial for maintaining homeostasis in the coupled cardio-respiratory systems. In the last several years, approaches proposing unidirectional and bidirectional quantities of relationship between cardiac cycles and respiratory rhythm have been increasingly used, with potential applicability in revealing new insights in many physiological and pathological conditions [1–10].

Many of these approaches work in the so-called framework of information dynamics, which provides measures able to assess the 'information content' of individual dynamic processes or collections of processes, and the information exchange among them [9,11–13]. A dynamical approach takes into account the flow of time by investigating how much the past system history contributes to reduce the uncertainty about the present state [11]. In this context, well-established tools like Granger Causality (*GC*) and Transfer Entropy (*TE*) assess pairwise directional interactions between two subsystems of a complex system, quantifying the directed flow of information from one subsystem to another [14–17]. On the other hand, measures of cross-predictability like cross-entropy (*CE*) quantify how much one process can be predicted from the other, returning asymmetric measures of coupling without implementing the Granger concept of causality [18]. All these measures have been widely applied in physiological contexts, e.g., for investigating cardiovascular variability [19,20], but also for studying cardiorespiratory interactions in healthy subjects [11] or in patients suffering of various types of sleep apnea [9].

In previous works, we investigated cardio-respiratory interactions in heart-failure patients using unidirectional markers and bidirectional markers based on linear models at baseline measurements [6,21]. The potential of the information-theoretic approaches described above has still not been tested for evaluating the effects of cardiac resynchronization therapy (CRT) in heart failure (HF). CRT has been used for more than 20 years in the treatment of symptomatic patients with heart failure with reduced left ventricular ejection fraction (LVEF) and abnormal QRS duration and morphology [22,23]. CRT leads to restoration of electromechanical synchrony, resulting in an improvement in the cardiac pump function, ejection fraction and reverse remodeling of the left ventricle. In most patients, these mechanical effects are associated with an improvement in the functional capacity and with a reduction in morbidity and mortality [24]. Nowadays, we know that, in addition to restoring ventricular mechanical synchronization, there are various cardiac and extra-cardiac effects of CRT, which are also responsible for its beneficial outcomes. One such effect is the remodeling of β-adrenergic signaling pathways and the restoration of sympathovagal balance [25]. The effect of CRT on cholinergic signaling has not been fully investigated in the literature yet, nor has the significance of CRT-induced neurohumoral modulation and its influence on cardio-respiratory interactions in these patients [26]. In recent years, significant advancements have been made in patient selection, implantation approaches, implant material performances, post-implantation device programming and medical therapy optimization. However, still a third of patients do not achieve the expected benefit from CRT implantation [27]. Therefore, it is very important—but also challenging—to use new approaches in the analysis of heart-failure patients who are candidates for CRT implantation, to preoperatively separate future responders from non-responders, according to parameters that have not yet been used. In this context, parameters that define autonomic function and cardio-respiratory interactions can certainly be of great importance.

In this work, we evaluate the cardio-respiratory interactions in terms of information flows between the coupled *RR* and respiratory signal dynamics. We used both the linear parametric and the nonlinear model-free implementations of the concept of Granger causality provided by *GC* and *TE* measures, as well as cross entropy, to investigate asymmetric coupling on different HF CRT responders and non-responders.

2. Materials and Methods

2.1. Subjects and Experimental Protocol

The data analyzed in this study consist in part of the raw data obtained from previously recruited patients with heart failure (reduced left ventricular ejection fraction, LVEF < 35%), and indication for CRT device implantation published in our previous work where the pool of results from all patients were analyzed together [21]. The study was approved by the Ethics Committee of the Faculty of Medicine the University of Belgrade, and each subject signed an informed consent form (Approve Date: 17 March 2017, Ref. Numb.29/III-4). Since the techniques applied in this work are sensitive to the type of heart rhythm, we excluded data from 8 patients (2 responders to CRT) with permanent atrial fibrillation from the group of 47 heart-failure patients. The main reason is that, since the cardiac rhythm in atrial fibrillation is highly irregular, the corresponding data would not satisfy stationarity criteria needed for our analysis; moreover, we have previously shown that the influence of respiration on cardiac rhythm and in the opposite direction is very small in patients with atrial fibrillation [6]. Contrary to our previous study [21] where we did not separate groups by arrhythmias, in this study we examined the effects of a low and high number of VESs on cardio-respiratory coupling before and after CRT. For this study, we divided the analyzed 39 HF patients into 3 groups: a first group of 14 patients with sinus rhythm (HFSin), a second group of 11 patients with a low number (<6) of ventricular extrasystoles (HFVES1), and a third group of 14 patients with 6 and more ventricular extrasystoles (HFVES2). Measurements were performed before (baseline) and approximately 9 months after CRT device implantation (follow-up). After follow-up, patients were divided into two groups, i.e., responders (N = 25) and non-responders (N = 14), in relation to the response to CRT, which was assessed according to changes in certain clinical and echocardiographic parameters. We defined an echocardiographic responder to CRT as a patient with increased left ventricular EF by 5% or more, and functional (clinical) who improved at least one class category in the New York Heart Association functional classification and/or the six-minute walk test by at least 10% at follow-up. In this study, a responder to resynchronization therapy had to meet both echocardiographic and functional criteria. Both baseline and follow-up experiments were conducted in the morning, between 7 and 8 a.m., in a quiet room surrounding at the Pacemaker Center of the University Clinical Center of Serbia. Baseline measurements were carried out immediately before device implantation. Data were acquired from 20 min of electrocardiographic (ECG) and respiratory (*Resp*)-signal measurements of relaxed subjects in the supine position and at a spontaneous breathing frequency by the Biopac MP100 system and AcqKnowledge 3.9.1 software (BIOPAC System, Inc., Santa Barbara, CA, USA) using a sampling rate equal to 1 kHz [21]. ECG data were acquired using the ECG 100C electrocardiogram amplifier module, while an RSP 100C respiratory pneumogram amplifier module with TSD 201 transducer attached to the belt (using an adjustable nylon strap) was used to measure abdominal expansion and contraction. The interbeat interval (i.e., ECG *RR* intervals) time series in patients with sinus rhythm and sinus rhythm with ventricular extrasystoles were analyzed. Interbeat (*RR*) intervals and interbreath (*BB*) intervals were extracted from signals recorded using the tool Pick Peaks from OriginPro 8.6 (OriginLab Corporation, Northampton, MA, USA). In the ECG with ventricular extrasystoles (VES), we used the time coordinate of the peak from VES as R peak coordinate for the subsequent analyses.

The first analysis consisted in evaluating the changes in *RR* and *BB* intervals by computing their mean and standard error. Furthermore, given that the *RR* and respiration samples obtained in this way were unequally positioned, an equal equidistant resampling of both series was carried out using the mean *RR* value of each individual as the resampling interval. The resampling procedure was performed via linear interpolation between two corresponding adjacent existing samples [3]. In this way, the resampling frequency was different for each subject, and fell within the range (0.8–1.5) Hz. The final analyzed *RR* and *Resp* time series were measured from signal lasting 20 min, resulting in a different numbers of samples according to the individually varying resampling frequency (Table 1).

Table 1. Comparison of the mean values of sample numbers in each group before (baseline) and after (follow-up) cardio resynchronization therapy (CRT) as well as of the mean values between the groups in each condition.

Group	Baseline ($\times 10^3$)	Follow-Up ($\times 10^3$)	p (Baseline vs. F-Up)
HFSin (N = 14; 71% R)	1.31 ± 0.18	1.22 ± 0.19	0.124
HFVES1 (N = 11; 64% R)	1.34 ± 0.29	1.25 ± 0.22	0.131
HFVES2 (N = 14; 57% R)	1.49 ± 0.18	1.30 ± 0.17	0.008

Table 1. *Cont.*

Group	Baseline ($\times10^3$)	Follow-Up ($\times10^3$)	p (Baseline vs. F-Up)
p (Among groups)	0.061	0.469	
Responders (N = 25)	1.43 ± 0.22	1.28 ± 0.18	0.007
Non-Responders (N = 14)	1.30 ± 0.22	1.22 ± 0.21	0.090
p (Resp. vs. Non-Resp.)	0.118	0.149	

Values are mean ± standard deviation. In the brackets percent of responders (R) in each HF group is given.

2.2. Information-Theoretic Measures and Data Analysis

For the analysis of *RR* and *BB* intervals, we computed the mean and standard error taking into account HFSin, HFVES1, HFVES2 groups, both at baseline and at follow-up for both responders and non-responders to CRT. Then, to investigate the dynamics of coupling and causality within the cardio-respiratory system, the information-theoretic measures of *GC*, *TE* and *CE* were computed on the *RR* and *Resp* time series.

These measures are defined, in the context of dynamic systems mapped by random processes, considering a bivariate random process $\mathbf{S} = \{X, Y\}$ composed of two interacting processes X and Y, respectively, considered as the driver and target process. In this context, the concept of Granger causality from X to Y is formalized through the measure of Transfer Entropy, which quantifies the information that the past states of driver process, $\mathbf{X}_n^- = [X_{n-1}, X_{n-2}, \ldots]$, transfer to the present state of the target process, Y_n, when the past states of the target itself, $\mathbf{Y}_n^- = [Y_{n-1}, Y_{n-2}, \ldots]$, are known. The *TE* is defined as:

$$TE_{X\to Y} = I\left(Y_n; \mathbf{X}_n^- \middle| \mathbf{Y}_n^-\right) \tag{1}$$

where $I\left(Y_n; \mathbf{X}_n^- \middle| \mathbf{Y}_n^-\right)$ is the conditional mutual information. On the other hand, the predictability of the present state of the target process Y_n from the past states of the driver process $\mathbf{X}_n^-$ is quantified through the cross entropy defined as:

$$CE_{X\to Y} = I\left(Y_n; \mathbf{X}_n^-\right) \tag{2}$$

where $I\left(Y_n; \mathbf{X}_n^-\right)$ is the mutual information.

In addition to the standard definitions given in (1) and (2), when assessing influences within physiological systems, it may be important to take into account the so-called instantaneous effects, i.e., the influence that the current state of driver process X_n can have on the target state Y_n [28,29]. To this end, the definitions of *TE* and *CE* can be modified as follows [28]:

$$iTE_{X\to Y} = I\left(Y_n; X_n, \mathbf{X}_n^- \middle| \mathbf{Y}_n^-\right) \tag{3}$$

$$iCE_{X\to Y} = I\left(Y_n; X_n, \mathbf{X}_n^-\right) \tag{4}$$

The so-called instantaneous *TE* and instantaneous *CE* defined in (3) and (4) are used when the instantaneous effects are deemed as physiologically relevant, i.e., in the direction from *Resp* to *RR* but not in the direction from *RR* to *Resp*.

In this work, the above presented measures were implemented in practice using both a linear model-based and a non-linear model-free estimator. The first approach makes use of linear regression models whereby the present state of the target process Y_n is described as a linear combination of its p past states, $\mathbf{Y}_n^p = \left[Y_{n-1}, \ldots, Y_{n-p}\right]$, or of the p past states of both processes, $\left[\mathbf{X}_n^p, \mathbf{Y}_n^p\right]$. These linear regression models return prediction errors, or residuals, whose variance can be related to the concept of conditional entropy under the assumption of Gaussianity and can be exploited to provide a measure of predictability improvement related to the *TE* [11]. Specifically, denoted as $\sigma^2_{Y_n|\mathbf{Y}_n^p}$, and $\sigma^2_{Y_n|\mathbf{X}_n^p,\mathbf{Y}_n^p}$ the prediction error variances of the two regression models, the measure of *GC* was computed as [30]

$$GC_{X\rightarrow Y} = \ln \frac{\sigma^2_{Y_n|\mathbf{Y}_n^p}}{\sigma^2_{Y_n|\mathbf{X}_n^p,\mathbf{Y}_n^p}} \tag{5}$$

Moreover, an extended measure of *GC* incorporating the instantaneous effect from *X* to *Y* and denoted as $iGC_{X\rightarrow Y}$ was computed, augmenting the vector $\mathbf{X}_n^p$ with the inclusion of X_n in (5) [31].

The second method is a model-free approach that makes use of nearest-neighbors to compute the probability density functions needed for the estimation of entropy measures [32]. Here, we used the formulation reported in [33] which performs, for each pattern forming a realization of the present and past states of the analyzed processes, the search for its *k* nearest neighbors to compute the entropy in the highest-dimensional space, and then adopts a distance-projection strategy to compute the entropies in the spaces of lower dimension. The resulting estimator has the form (see [34] for details)

$$TE_{X\rightarrow Y} = \psi(k) + \left\langle \psi\left(N_{\mathbf{Y}_n^q}+1\right) - \psi\left(N_{Y_n\mathbf{Y}_n^q}+1\right) - \psi\left(N_{\mathbf{X}_n^q\mathbf{Y}_n^q}+1\right)\right\rangle \tag{6}$$

where $\psi(\cdot)$ is the digamma function, *k* is the number of neighbors considered in the highest dimensional space, $\varepsilon_{n,k}$ is twice the distance of the *n*-th reference pattern $\left(y_n, \mathbf{y}_n^q, \mathbf{x}_n^q\right)$ to its *k*th neighbor in the highest dimensional space, and $N_{\mathbf{X}_n^q\mathbf{Y}_n^q}$, $N_{Y_n\mathbf{Y}_n^q}$ and $N_{\mathbf{Y}_n^q}$ are the number of patterns whose distance from the projected reference pattern ($(\mathbf{y}_n^q, \mathbf{x}_n^q)$, $(y_n, \mathbf{y}_n^q)$ and $\mathbf{y}_n^q$, respectively) is lower than $\varepsilon_{n,k}/2$. A similar treatment leads to estimating the cross entropy measure as:

$$CE_{X\rightarrow Y} = \psi(k) + \psi(N) - \left\langle \psi(N_{Y_n}+1) + \psi\left(N_{\mathbf{X}_n^q}+1\right)\right\rangle \tag{7}$$

where *N* is the number of available patterns and the other symbols have the same meaning as in (6) (thus $N_{\mathbf{X}_n^q}$ and N_{Y_n} are the number of patterns whose distance from the projected $\mathbf{x}_n^q$ and y_n is lower than $\varepsilon_{n,k}/2$). As for the linear estimation of *GC*, extended measures of *TE* and *CE* were also obtained augmenting the vector $\mathbf{X}_n^q$ with the inclusion of X_n in (6) and (7) when the instantaneous effect from *X* to *Y* was deemed as causally relevant.

In the analyzed dataset, the measures defined above were computed along the two directions of interaction from *Resp* to *RR* and from *RR* to *Resp*, to analyze closed-loop bidirectional cardiorespiratory interactions. Given the adopted measurement convention, the heartbeat cannot transfer information at zero-lag to the respiratory system, since the *i*-th breath sample is simultaneous with the onset of the *i*-th cardiac period. Therefore, the standard measures of *GC*, *TE* and *CE* given in (5)–(7) computed without incorporating the instantaneous effect were used when *X* = *RR* and *Y* = *Resp*, while the extended measures in which the vector of the past driver states is augmented with the present driver state were used when *X* = *Resp* and *Y* = *RR*.

With regard to *GC* computation, the model order *p* was set according to the Akaike information criterion (AIC), taking care to consider the present state of the driver in defining the model in the direction of analysis from *Resp* to *RR*. Even if the choice of the model order can influence the value of the achieved information measures and thus their physiological significance, in this work a sufficiently high value was selected in order to have enough 'memory' of the two processes involved in the interaction [34].

Regarding the non-linear implementation of *CE* and *TE* measures, the number of neighbors *k* was fixed to 10 and the number of considered past states *q* to 2, as commonly performed in the literature in the field [35].

2.3. Statistical Analysis

In order to assess the statistical significance of the differences of the various indexes between groups and conditions, we employed nonparametric tests. For each measure, the distributions of the results were tested against normality through a Shapiro–Wilk test and since the most of the measures did not follow a normal distribution, we performed nonparametric tests for their comparison.

We performed the Wilcoxon test to compare the mean values of sample numbers in each group before and after CRT, and the Mann–Whitney U test to compare CRT groups in each condition. The Kruskal–Wallis H test was applied to compare the mean values among the HF groups (HFSin, HFVES1 and HFVES2) in each condition (Table 1).

For *RR* and *BB* intervals analyses we tested (1) the baseline versus follow-up condition using the Wilcoxon test for each HF group, and the CRT group; (2) responders versus non-responders performing the Mann–Whitney U test in both conditions; and (3) differences between the HF groups, before and after CRT, using also the Mann–Whitney U test ($n = 3$: HFSin vs. HFVES1, HFSin vs. HFVES2, and HFVES1 vs. HFVES2), Table 2.

Table 2. Descriptive statistics.

		HF Groups			CRT Groups	
	Condition	**HFSin**	**HFVES1**	**HFVES2**	**Responders**	**Non-Responders**
RR [s]	Baseline	0.930 ± 0.032 $^{\tau\tau}$	0.932 ± 0.054	0.813 ± 0.026 **	0.857 ± 0.027 *	0.943 ± 0.040
	Follow-up	1.010 ± 0.053	0.99± 0.40	0.934 ± 0.031	0.959 ± 0.036	1.007 ± 0.051
BB [s]	Baseline	4.25 ± 0.38	3.82 ± 0.27	3.40 ± 0.13	4.01 ± 0.24 *	3.49 ± 0.14
	Follow-up	4.30 ± 0.39	3.96 ± 0.37	3.65 ± 0.21	4.32 ± 0.26 #	3.35 ± 0.14

** $p < 0.01$ * $p < 0.05$ Baseline vs. Follow-up, # $p < 0.05$ Responders vs. Non-Responders, $^{\tau\tau}$ $p < 0.01$ HFSin vs. HFVES2.

For information-theoretic measures, the following statistical analyses were performed in HF groups and in CRT groups separately. We used the Mann–Whitney U test to compare groups at baseline and at follow-up measurements. In each group, significant differences between bidirectional measures distributions were recognized by a Wilcoxon non-parametric pairwise test. For comparison of these parameters between baseline and follow-up measurements, we also used a Wilcoxon test.

For all analyses, probability values $p < 0.05$ were considered statistically significant. Statistical analyses were carried out in SPSS (SPSS Inc., Chicago, IL, USA) version 17.

3. Results

In Table 2, the analysis of *RR* intervals and *BB* intervals in the analyzed groups is shown, with regard to descriptive statistics measures (mean and standard error). Responders to CRT exhibited statistically significantly prolonged *RR* intervals ($p = 0.028$) and *BB* intervals ($p = 0.014$) compared to baseline, while non-responders did not change these markers ($p > 0.05$). Also, there was no difference between the responder and non-responder groups, except for *BB* intervals at follow-up measurements ($p = 0.013$).

Figures 1–3 present, as bar graphs, the results obtained with regard to linear and nonlinear measures of *GC*, *TE*, and *CE* computed along the two directions of interaction from *Resp* to *RR* and from *RR* to *Resp*. There was a tendency for *GC*(*Resp-RR*) to be higher than *GC*(*RR-Resp*) in all patients at baseline condition, but only after CRT device implantation in the HFSin and HFVES2 group the difference was statistically significant, with $p < 0.05$ (Figure 1).

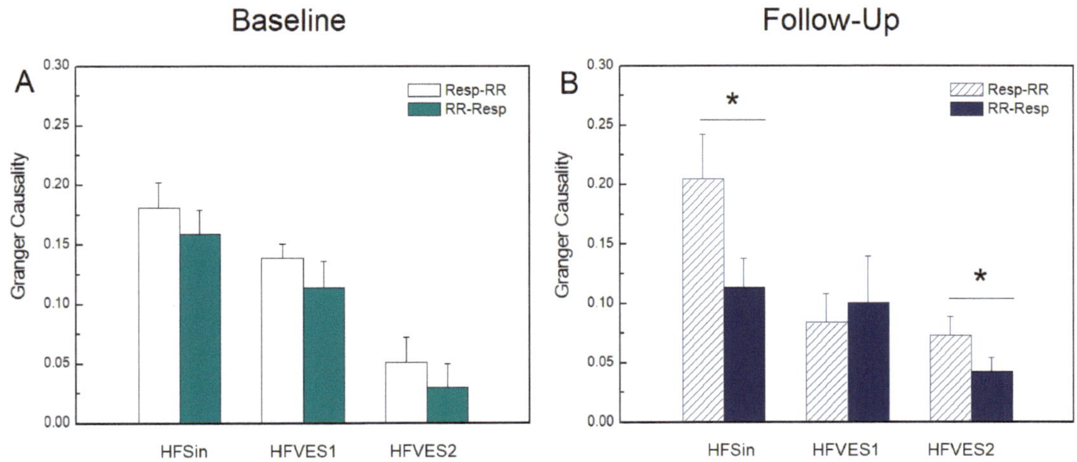

Figure 1. Granger causality (*GC*) at baseline measurements (**A**) and at follow-up measurements after CRT device implantation (**B**) in heart-failure patients: with sinus rhythm (HFSin), with sinus rhythm and small number of ventricular extrasystoles (HFVES1), and with sinus rhythm and higher number of ventricular extrasystoles (HFVES2). *GC(Resp-RR)* and *GC(RR-Resp)* denote causality of respiration in *RR* intervals time series and vice versa. Data are presented as mean + standard error. * $p < 0.05$.

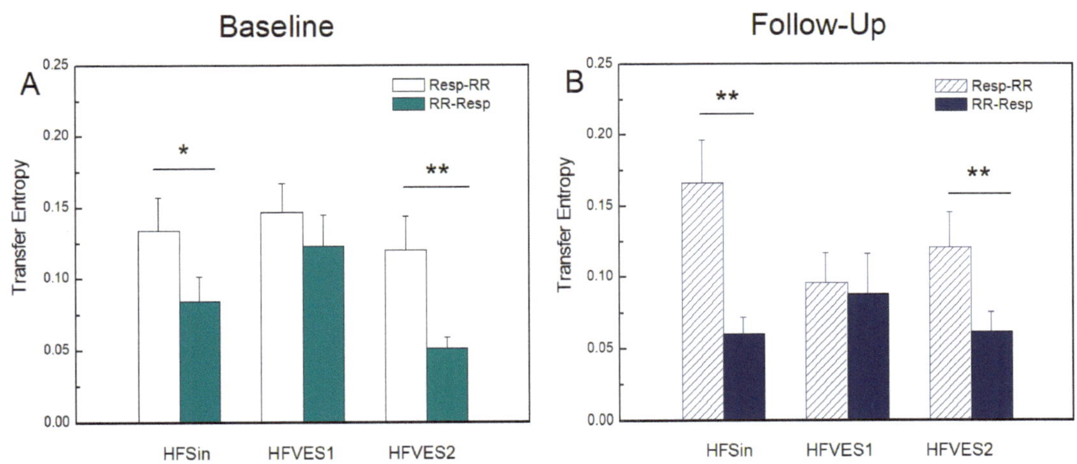

Figure 2. Transfer entropy (*TE*) at baseline measurements (**A**) and at follow-up measurements after CRT device implantation (**B**) in heart-failure patients: with sinus rhythm (HFSin), with sinus rhythm and small number of ventricular extrasystoles (HFVES1), and with sinus rhythm and higher number of ventricular extrasystoles (HFVES2). *TE(Resp-RR)* and *GC(RR-Resp)* denote causality of respiration in RR intervals time series and vice versa. Data are presented as mean + standard error. ** $p < 0.01$, * $p < 0.05$.

Moreover, compared with the baseline measurement, in the group of HFSin patients, CRT showed significantly decreased *GC(RR-Resp)*, $p = 0.048$ (not shown in Figure 1), resulting in a significant difference between *GC(Resp-RR)* and *GC(RR-Resp)* ($p < 0.05$) (Figure 1B). In the group of HFVES1 patients, there was no statistically significant difference between *GC* in both measurements (Figure 1).

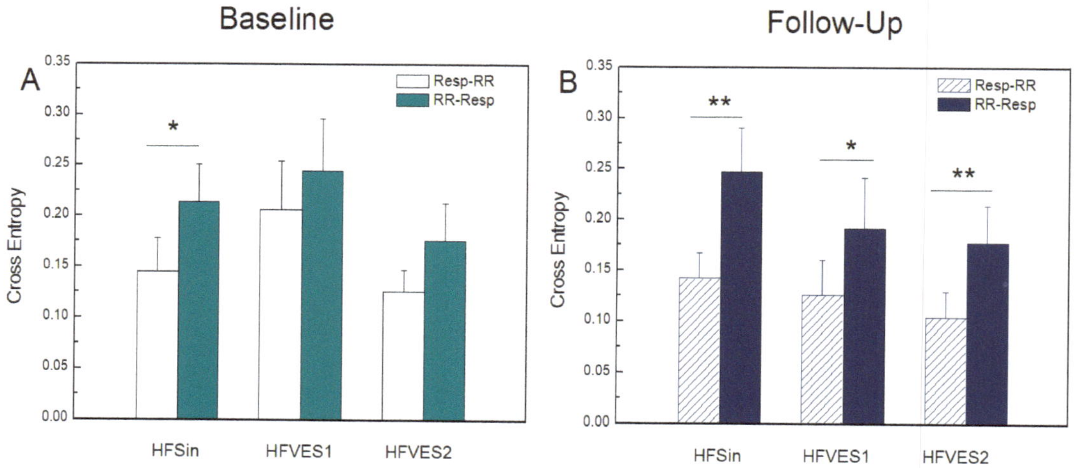

Figure 3. Cross entropy (*CE*) at baseline measurements (**A**) and at follow-up measurements after CRT device implantation (**B**) in heart failure patients: with sinus rhythm (HFSin), with sinus rhythm and small number of ventricular extrasystoles (HFVES1), and with sinus rhythm and higher number of ventricular extrasystoles (HFVES2). *TE*(*Resp-RR*) and *CE*(*RR-Resp*) denote causality of respiration in RR intervals time series and vice versa. Data are presented as mean + standard error. ** $p < 0.01$, * $p < 0.05$.

In Table 3, the results of the comparisons between coupling and causality measures computed in the HF groups at baseline and at follow-up measurements are presented. Linear Granger causality measures were sensitive to a higher number of VESs and at baseline measurements they significantly decreased with the number of VESs in both directions. After CRT implementation, a statistically significant difference was found in *GC*(*Resp-RR*) between the HFSin patients' group and the two other groups and in *GC*(*RR-Resp*) between the HFSin and the HFVES2 group.

Table 3. Statistical significance of comparisons between pairs of groups for bidirectional measures.

	Baseline			Follow-Up		
	HFSin vs. HFVES1	HFSin vs. HFVES2	HFVES1 vs. HFVES2	HFSin vs. HFVES1	HFSin vs. HFVES2	HFVES1 vs. HFVES2
GC(*Resp-RR*)	0.432	**0.001**	**0.002**	**0.013**	**0.004**	0.751
GC(*RR-Resp*)	0.297	**0.001**	**0.001**	0.212	**0.012**	0.527
TE(*Resp-RR*)	0.572	0.635	0.165	0.080	0.246	0.681
TE(*RR-Resp*)	0.181	0.137	**0.005**	1.000	0.946	0.918
CE(*Resp-RR*)	0.258	0.131	0.123	0.440	0.781	0.758
CE(*RR-Resp*)	0.643	0.274	0.258	0.411	0.980	0.957

Statistically significant values are in bold.

Figure 2 shows the measures of *TE* computed along the two directions of interaction between *Resp* and *RR*. A statistically significant difference between *TE(Resp-RR)* and *TE*(*RR-Resp*) in the HFSin and HFVES2 groups of patients was present at the baseline and was maintained after CRT implantation ($p < 0.01$), also showing a tendency to become more marked. Only in the HFVES1 group, *TE*(*Resp-RR*) was not significantly greater than *TE*(*RR-Resp*). In the comparison between groups at the baseline measurement, we found a statistically significant difference in *TE*(*RR-Resp*) between groups of patients with VES (Table 3). CRT influenced *TE*(*RR-Resp*) in the HFSin and HFVES1 group with decreasing values but only towards to the tendency of statistical significance ($p = 0.08$) and ($p = 0.12$).

Contrary to *GC* and *TE*, where *RR* to *Resp* measures were lower than *Resp* to *RR* measures, *CE*(*RR*-*Resp*) were higher than *CE*(*Resp*-*RR*) in all analyzed groups (Figure 3). Statistically significant differences were preserved only in HFSin group during both baseline and follow-up. In the other two groups, statistically significant differences were reported only after CRT implantation. Comparison between groups did not reveal any statistically significant difference in *CE* measures.

Figure 4 presents the results of the comparison between baseline and follow-up markers and between the two directions, in both groups of responders and non-responders. In the group of CRT responders, the two *GC* measures did not change after CRT, while in the group of non-responders, follow-up measurements indicated a reduction in *GC*(*RR*-*Resp*). In this group, although unchanged, *GC*(*Resp*-*RR*) became significantly higher than *GC*(*RR*-*Resp*) after the follow-up measurement. Further, *TE*(*Resp*-*RR*) was significantly higher than *TE*(*RR*-*Resp*) in both baseline and follow-up measurements in responders, and in non-responders only after follow-up measurement. Contrary, *CE*(*Resp*-*RR*) was significantly lower than *CE*(*RR*-*Resp*) in both measurements in responders and in non-responders only after follow-up measurement. No significant differences between directed measures for any measure were found at baseline in non-responders while significant differences were obtained for all information-theoretic measures at baseline in the group of CRT responders.

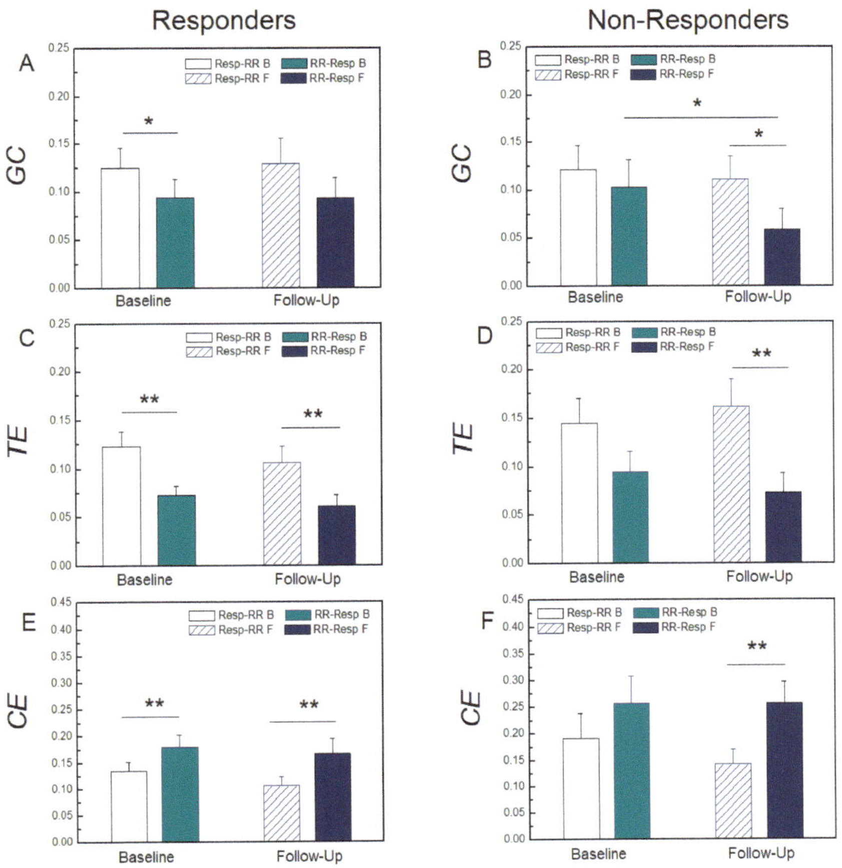

Figure 4. Granger causality (*GC*) at baseline measurements and at follow-up measurements in responders (**A**) and non-responders (**B**) to CRT. Transfer Entropy (*TE*) at the baseline measurements

and at the follow-up measurements in responders (**C**) and non-responders (**D**) to CRT. Cross Entropy (*CE*) at the baseline measurements and at the follow-up measurements in responders (**E**) and non-responders (**F**) to CRT. *Resp-RR* and *RR-Resp* denote causality of respiration in RR interval time series and vice versa. Data are presented as mean + standard error. ** $p < 0.01$, * $p < 0.05$.

4. Discussion

4.1. Heart Rate and Respiratory Rate

The analysis of the mean heart and respiratory rates confirms the results from previous works, and here it is complemented with a dynamic analysis of the variability of *RR* and *BB* intervals [6]. The results of our research indicate that in patients with heart failure, resting heart and respiratory rates are the highest in patients with a large number of ventricular extrasystoles before CRT. When we separately analyze CRT responders and non-responders, we observe that during follow-up there is a decrease in heart rate in all HF patients, but that it is statistically significant only in CRT responders. The reasons for this result are related to the intensification of antiarrhythmic therapy in all patients after CRT implantation, as well as to the improvement in clinical status and the recovery of vagal tone, which dominantly determines this parameter, in patients who benefited from device implantation [36,37].

Respiratory rate also significantly decreases in CRT responders during follow-up, but increases slightly in non-responders, resulting in a statistically significant difference in the values of this parameter between these two groups at control examination. An improvement in cardiac function enables better perfusion, i.e., a reduction in tissue hypoxia, which leads to reduction in chemoreceptor stimulation, and this can explain the decrease in respiratory rate in CRT responders [38]. Our results also confirm that better functional capacity, according to the NYHA (New York Heart Association) classification of HF, and higher LVEF, are associated with lower respiratory frequency [39].

4.2. Granger Causality and Transfer Entropy

In this study the model-based Granger causality analysis showed a stronger influence of respiration on heart rhythm in all HF patients. The reasons for small discrepancies in *GC* analysis comparing these results and the findings reported in our previous study in the HF groups are probably methodological, since we used a different *GC* algorithm in the last paper [40]. In addition, the groups with VES were not equally defined: in our previous work, the HFVES group was defined by more than 20 premature ventricular ectopic beats during signal recording [6]. According to Granger causal analyses, such as the *GC* measure (model-based) and the *TE* measure (model-free), it is expected that respiration would have a stronger impact on the heart rate than vice versa. This result can be documented in healthy subjects and describes the well-known phenomenon of respiratory sinus arrhythmia (RSA) whereby *Resp* changes modulate the heart period duration differently during inspiration than during expiration. Our results, for the HF patients in sinus rhythm, document this effect only when the *TE* was used (Figure 2A), but not using the *GC* (Figure 1A), suggesting that RSA mechanisms are nonlinear and cannot be fully captured using the linear *GC* analysis.

Furthermore, the *GC* decreased progressively from sinus rhythm to the HFVES1 and the HFVES2 group during baseline (Figure 1A), while the *TE* did not show an evident similar trend (Figure 2A). This suggests that in HF patients, the cardiorespiratory interactions evaluated in the presence of extrasystoles have a strong nonlinear component which can be properly captured only when *TE* is estimated.

A similar influence of cardiac rhythm on respiration and vice versa in the HFVES1 group at the control recording is difficult to interpret. The effect of respiration on cardiac rhythm is determined by various factors even under physiological conditions. For instance, it has been shown that this influence loses its importance in the elderly, i.e., the strength of direct respiratory modulation of cardiac rhythm decreases with aging, as well as the indirect effects of respiration on heart rate due to the reduction in baroreflex sensitivity [41–44]. When autonomic imbalance occurs (as in HF), this bidirectional interaction is certainly altered, with the activation of compensatory mechanisms that are still poorly understood. In the previous paper [6], we presented the assumption that the stronger influence of the heart rhythm on the respiratory signal in patients with HF and VES is due to compensatory mechanisms that restore the regularity of the heart rhythm and increase its influence on breathing. However, after this study, it is clear that the frequency of extrasystoles, as well as possible other factors such as the HF stage or age, have an impact on bidirectional cardio-respiratory interactions. On the other hand, we observed that during follow-up, the influence of breathing on heart rhythm significantly weakened in patients with a low number of VESs, thus demonstrating a stronger influence of the cardiac rhythm on the respiratory signal in the HFVES1 group.

Regarding the *TE* values in HFVES patients, we noticed that causality along the direction from the *RR* to *Resp* in the follow up measurements was not sensitive to the higher frequency of extrasystoles, as it was in baseline recordings (Table 3). On the other hand, the prevalence of the causal interaction along the direction *Resp* to *RR* becomes much more marked after CRT, and this was detected using both *GC* and *TE* (Figures 1B and 2B). This suggests that resynchronization therapy is successful in restoring RSA in HF patients, activating both the linear and nonlinear components of the interaction from *Resp* to *RR*. The recovery of vagal tone and baroreceptor activity is responsible for strengthening the effect of respiration on cardiac rhythm, i.e., that intense cardiorespiratory interaction in this direction is a confirmation of the existence of the balanced activity of control and compensatory mechanisms. This result also implies that with the restoration of *Resp* on *RR*, we can detect the joined mechanism of the reduced linear component of cardiac causality from *RR* on restored *Resp* (Figure 1A,B).

4.3. Cross Entropy

The opposite trends exhibited by *CE* if compared to *GC* and *TE* reflect a prevalence of the coupling in the direction *RR* to *Resp*, amplified after CRT (Figure 3). The reason for this difference is very likely methodological: while *GC* and *TE* quantify directed interactions in terms of predictability improvement, the *CE* implements the concept of cross-predictability. From this point of view, the *CE* is more similar to measures like convergent cross-mapping, which attempt to infer the dominant causation in a bidirectional interaction by predicting the driver starting from the history of the target process [45]. Therefore, according to this methodological consideration, it is not surprising that a predominant *Resp* to *RR* mechanism produces a higher *GC*/*TE* from *Resp* to *RR* but also higher *CE* from *RR* to *Resp*.

The values of *CE*(*Resp-RR*) and *CE*(*RR-Resp*) were not significantly different between the groups at any of the measurements, and both parameters had the largest preoperative values in the HFVES1 group, and at control recording in the HFSin group. This suggests that in cardiorespiratory coupling, the prediction of the dynamics of one signal based solely on the history of the other is stable in time and that the presence of arrhythmias does not significantly affect it.

Our finding that in HF patients, cardio-respiratory interactions assessed in the presence of ventricular extrasystoles have a strong nonlinear component that can only be properly captured when using *TE*, was confirmed by the results of model-free *CE*, which did not exhibit any particular trends, and after the resynchronization therapy clearly differentiated the two directions of interaction even in presence of extrasystoles. Transfer entropy, in addition to confirming that the occurrence of ventricular extrasystoles significantly affects the interactions of the respiratory and cardiovascular system, also indicates that their frequency decisively determines the intensity of information flow between these systems. What compensatory mechanisms and at what levels of regulation of physiological processes trigger the occurrence of ventricular extrasystoles, and what are the causes of their pronounced dynamics during monitoring, as well as in relation to the extra beats frequency, remain to be determined by future research.

4.4. Cardiac Resynchronization Therapy

The results of the comparative analysis of CRT responders and non-responders provide useful insights about the reaction of patients to therapy. At the baseline measurement, no significant differences between these two groups were reported with regard to Granger causality analysis, with a stronger influence of respiration on heart rhythm in both groups of patients at the control examination in the case of the model-free approach (Figure 4C,D) and only for non-responders with the linear measure (Figure 4B).

Moreover, it is noteworthy that the nonlinear measure of causality is able to identify the prevalence of causality from respiration to cardiac activity in the baseline condition for responders to CRT (Figure 4C). This finding may suggest that if there is a statistically significant difference between *TE* parameters in a patient who is a candidate for CRT, he/she will benefit from device implantation. However, this would be too bold and probably a wrong conclusion, given that in CRT non-responders, the difference between these parameters preoperatively is close to statistical significance, and is achieved at the follow-up examination.

The values of both CE parameters were lower in CRT responders compared to non-responders for all the measurements. During follow-up, statistically significantly higher values of *CE*(*RR-Resp*) compared to *CE*(*Resp-RR*) were observed in CRT responders and non-responders.

It is also very interesting to observe changes in the intensity of the information flow in different directions in CRT non-responders. In fact, the *TE*(*Resp-RR*) increase in patients with significantly reduced vagal tone is an indicator that the values of RSA, which is a measure of cardiorespiratory interaction, and *CE*(*Resp-RR*), which is a measure of cardio-respiratory coupling, do not necessarily follow each other, and this is emphasized in some previous works [46].

5. Conclusions

Information-theoretic analysis of bidirectional cardio-respiratory interactions revealed the strong effect of nonlinear components in the *RR* and *Resp* time series, as well as in their causal interaction in evaluation of cardiac resynchronization therapy in HF patients.

Our results, for the HF patients in sinus rhythm, document the RSA effect only when using the *TE* but not *GC*. On the other hand, the prevalence of the causal interaction along the direction from *Resp* to *RR* becomes much more marked after CRT, and this was detected using both *GC* and *TE*, which suggests that the resynchronization therapy is successful in restoring RSA in HFSin patients, activating both the linear and nonlinear components of the interaction from *Resp* to *RR*. Regarding the *CE*, it exhibits opposite trends of *GC* and *TE*, showing prevalence of coupling in the direction from *RR* to *Resp*, enhanced after CRT.

In the HF patients, before and after CRT, the occurrence of a high number of extrasystoles significantly reduced *GC* in both directions. The *GC* progressively decreases from HF in sinus rhythm to HFVES1 and to HFVES2, while the *TE* does not show a similar evident trend. This finding suggests that in HF patients, the cardio-respiratory interactions evaluated in the presence of extrasystoles have a strong nonlinear component that can only be properly captured when the *TE* is used. This is confirmed also by the *CE* formulated in a model-free way, which does not show particular trends and after the resynchronization therapy clearly differentiates the two directions of interaction even in the presence of extrasystoles.

Contrary to results obtained in the group of non-responders, in the group of responders to CRT we found a statistically significant difference in the baseline condition between the *Resp-RR* and *RR-Resp* directions, quantified by applied measures from information theory. This finding could be useful in the selection of patients for CRT.

Author Contributions: Conceptualization, M.M.P., N.N.R., S.U.P. and L.F.; methodology, L.F., R.P., C.B. and M.M.P.; formal analysis, M.M.P.; data curation, N.N.R. and S.U.P.; writing—original draft preparation, M.M.P., N.N.R., R.P., C.B. and L.F. All authors have read and agreed to the published version of the manuscript.

Funding: This research received no external funding.

Institutional Review Board Statement: The study was conducted in accordance with the Declaration of Helsinki, and approved by the Ethics Committee of the Faculty of Medicine the University of Belgrade, and each subject signed an informed consent form (Approve Date: 17 March 2017, Ref. Numb. 29/III-4).

Data Availability Statement: Data will be made available on request.

Acknowledgments: M.P. was partially financed by Ministry for Science, Technological Development and Innovation of the Republic Serbia [200110]. This study was supported by SiciliAn MicronanOTecH Research and Innovation CEnter "SAMOTHRACE" (MUR, PNRR-M4C2, ECS_00000022), spoke 3—Università degli Studi di Palermo "S2-COMMs—Micro and Nanotechnologies for Smart & Sustainable Communities". R.P. was partially supported by the European Social Fund (ESF)—Complementary Operational Programme (POC) 2014/2020 of the Sicily Region.

Conflicts of Interest: The authors declare no conflict of interest.

References

1. Porta, A.; Guzzetti, S.; Montano, N.; Pagani, M.; Somers, V.; Malliani, A.; Baselli, G.; Cerutti, S. Information domain analysis of cardiovascular variability signals: Evaluation of regularity, synchronisation and co-ordination. *Med. Biol. Eng. Comput.* **2000**, *38*, 180–188. [CrossRef] [PubMed]
2. de Abreu, R.M.; Cairo, B.; Porta, A. On the significance of estimating cardiorespiratory coupling strength in sports medicine. *Front. Netw. Physiol.* **2023**, *2*, 1114733. [CrossRef]
3. Kapidžić, A.; Platiša, M.M.; Bojić, T.; Kalauzi, A. Nonlinear properties of cardiac rhythm and respiratory signal under paced breathing in young and middle-aged healthy subjects. *Med. Eng. Phys.* **2014**, *36*, 1577–1584. [CrossRef] [PubMed]
4. Borovkova, E.I.; Prokhorov, M.D.; Kiselev, A.R.; Hramkov, A.N.; Mironov, S.A.; Agaltsov, M.V.; Ponomarenko, V.I.; Karavaev, A.S.; Drapkina, O.M.; Penzel, T. Directional couplings between the respiration and parasympathetic control of the heart rate during sleep and wakefulness in healthy subjects at different ages. *Front. Netw. Physiol.* **2022**, *2*, 942700. [CrossRef] [PubMed]
5. Sorelli, M.; Hutson, T.N.; Iasemidis, L.; Bocchi, L. Linear and Nonlinear Directed Connectivity Analysis of the Cardio-Respiratory System in Type 1 Diabetes. *Front. Netw. Physiol.* **2022**, *2*, 840829. [CrossRef]
6. Radovanović, N.N.; Pavlović, S.U.; Milasinovic, G.; Kircanski, B.; Platiša, M. Bidirectional Cardio-Respiratory Interactions in Heart Failure. *Front. Physiol.* **2018**, *9*, 165. [CrossRef]
7. Elstad, M.; O'callaghan, E.L.; Smith, A.J.; Ben-Tal, A.; Ramchandra, R. Cardiorespiratory interactions in humans and animals: Rhythms for life. *Am. J. Physiol. Circ. Physiol.* **2018**, *315*, H6–H17. [CrossRef]
8. Platiša, M.M.; Radovanović, N.N.; Kalauzi, A.; Milašinović, G.; Pavlović, S.U. Multiscale Entropy Analysis: Application to Cardio-Respiratory Coupling. *Entropy* **2020**, *22*, 1042. [CrossRef]
9. Lazic, I.; Pernice, R.; Loncar-Turukalo, T.; Mijatovic, G.; Faes, L. Assessment of Cardiorespiratory Interactions during Apneic Events in Sleep via Fuzzy Kernel Measures of Information Dynamics. *Entropy* **2021**, *23*, 698. [CrossRef]
10. Kalauzi, A.; Matić, Z.; Platiša, M.M.; Bojić, T. Two Operational Modes of Cardio-Respiratory Coupling Revealed by Pulse-Respiration Quotient. *Bioengineering* **2023**, *10*, 180. [CrossRef]
11. Faes, L.; Porta, A.; Nollo, G. Information Decomposition in Bivariate Systems: Theory and Application to Cardiorespiratory Dynamics. *Entropy* **2015**, *17*, 277–303. [CrossRef]
12. Xiong, W.; Faes, L.; Ivanov, P.C. Entropy measures, entropy estimators, and their performance in quantifying complex dynamics: Effects of artifacts, nonstationarity, and long-range correlations. *Phys. Rev. E* **2017**, *95*, 062114. [CrossRef]
13. Faes, L.; Porta, A.; Nollo, G.; Javorka, M. Information Decomposition in Multivariate Systems: Definitions, Implementation and Application to Cardiovascular Networks. *Entropy* **2016**, *19*, 5. [CrossRef]
14. Granger, C.W.J. Investigating causal relations by econometric models and cross-spectral methods. *Econom. J. Econom. Soc.* **1969**, *37*, 424–438. [CrossRef]
15. Barrett, A.B.; Barnett, L.; Seth, A.K. Multivariate Granger causality and generalized variance. *Phys. Rev. E Stat. Nonlin. Soft Matter Phys.* **2010**, *81 Pt 1*, 041907. [CrossRef]
16. Porta, A.; Faes, L. Wiener–Granger Causality in Network Physiology with Applications to Cardiovascular Control and Neuroscience. *Proc. IEEE* **2015**, *104*, 282–309. [CrossRef]
17. Pernice, R.; Faes, L.; Feucht, M.; Benninger, F.; Mangione, S.; Schiecke, K. Pairwise and higher-order measures of brain-heart interactions in children with temporal lobe epilepsy. *J. Neural Eng.* **2022**, *19*, 045002. [CrossRef]
18. Faes, L.; Nollo, G.; Porta, A. Information Decomposition: A Tool to Dissect Cardiovascular and Cardiorespiratory Complexity. In *Complexity and Nonlinearity in Cardiovascular Signals*; Springer: Cham, Switzerland, 2017; pp. 87–113. [CrossRef]
19. Faes, L.; Porta, A. Conditional Entropy-Based Evaluation of Information Dynamics in Physiological Systems. In *Directed Information Measures in Neuroscience*; Springer: Berlin/Heidelberg, Germany, 2014; pp. 61–86. [CrossRef]
20. Porta, A.; Faes, L.; Nollo, G.; Bari, V.; Marchi, A.; De Maria, B.; Takahashi, A.C.M.; Catai, A.M. Conditional Self-Entropy and Conditional Joint Transfer Entropy in Heart Period Variability during Graded Postural Challenge. *PLoS ONE* **2015**, *10*, e0132851. [CrossRef]

21. Radovanović, N.N.; Pavlović, S.U.; Milašinović, G.; Platiša, M.M. Effects of Cardiac Resynchronization Therapy on Cardio-Respiratory Coupling. *Entropy* **2021**, *23*, 1126. [CrossRef]
22. Leyva, F.; Nisam, S.; Auricchio, A. 20 Years of Cardiac Resynchronization Therapy. *J. Am. Coll. Cardiol.* **2014**, *64*, 1047–1058. [CrossRef]
23. Glikson, M.; Nielsen, J.C.; Kronborg, M.B.; Michowitz, Y.; Auricchio, A.; Barbash, I.M.; Barrabés, J.A.; Boriani, G.; Braunschweig, F.; Brignole, M.; et al. 2021 ESC Guidelines on cardiac pacing and cardiac resynchronization therapy. *Eur. Heart J.* **2021**, *42*, 3427–3520. [CrossRef] [PubMed]
24. Dhesi, S.; Lockwood, E.; Sandhu, R.K. Troubleshooting Cardiac Resynchronization Therapy in Nonresponders. *Can. J. Cardiol.* **2017**, *33*, 1060–1065. [CrossRef] [PubMed]
25. Chakir, K.; Daya, S.K.; Aiba, T.; Tunin, R.S.; Dimaano, V.L.; Abraham, T.P.; Jaques-Robinson, K.M.; Lai, E.W.; Pacak, K.; Zhu, W.Z.; et al. Mechanisms of enhanced beta-adrenergic reserve from cardiac resynchronization therapy. *Circulation* **2009**, *119*, 1231–1240. [CrossRef] [PubMed]
26. DeMazumder, D.; Kass, D.A.; O'rourke, B.; Tomaselli, G.F. Cardiac Resynchronization Therapy Restores Sympathovagal Balance in the Failing Heart by Differential Remodeling of Cholinergic Signaling. *Circ. Res.* **2015**, *116*, 1691–1699. [CrossRef] [PubMed]
27. Daubert, C.; Behar, N.; Martins, R.P.; Mabo, P.; Leclercq, C. Avoiding non-responders to cardiac resynchronization therapy: A prac-tical guide. *Eur. Heart J.* **2017**, *38*, 1463–1472.
28. Faes, L.; Nollo, G.; Porta, A. Compensated Transfer Entropy as a Tool for Reliably Estimating Information Transfer in Physiological Time Series. *Entropy* **2013**, *15*, 198–219. [CrossRef]
29. Pernice, R.; Lazic, I.; Bara, C.; Sparacino, L.; Mijatovic, G.; Loncar-Turukalo, T.; Faes, L. Assessment of Cardiorespiratory Interactions During Spontaneous and Controlled Breathing: Non-linear Model-free Analysis. In Proceedings of the 2022 12th Conference of the European Study Group on Cardiovascular Oscillations (ESGCO), Štrbské Pleso, Slovakia, 9–12 October 2022; pp. 1–2. [CrossRef]
30. Geweke, J. Measurement of linear dependence and feedback between multiple time series. *J. Am. Stat. Asso-Ciation* **1982**, *77*, 304–313. [CrossRef]
31. Schiatti, L.; Nollo, G.; Rossato, G.; Faes, L. Extended Granger causality: A new tool to identify the structure of physiological networks. *Physiol. Meas.* **2015**, *36*, 827–843. [CrossRef]
32. Kozachenko, L.F.; Leonenko, N.N. Sample Estimate of the Entropy of a Random Vector. *Probl. Peredachi Inf.* **1987**, *23*, 9–16.
33. Kraskov, A.; Stögbauer, H.; Grassberger, P. Estimating mutual information. *Phys. Rev. E Stat. Nonlin. Soft Matter Phys.* **2004**, *69*, 066138; Erratum in **2011**, *83*, 019903. [CrossRef]
34. Faes, L.; Erla, S.; Nollo, G. Measuring Connectivity in Linear Multivariate Processes: Definitions, Interpretation, and Practical Analysis. *Comput. Math. Methods Med.* **2012**, *2012*, 140513. [CrossRef]
35. Faes, L.; Kugiumtzis, D.; Nollo, G.; Jurysta, F.; Marinazzo, D. Estimating the decomposition of predictive information in multivariate systems. *Phys. Rev. E Stat. Nonlin. Soft Matter Phys.* **2015**, *91*, 032904. [CrossRef]
36. Alvarez-Alvarez, B.; García-Seara, J.; Martínez-Sande, J.L.; Rodríguez-Mañero, M.; López, X.A.F.; González-Melchor, L.; Bermejo, R.M.A.; Iglesias-Alvarez, D.; Sampedro, F.G.; Díaz-Louzao, C.; et al. Cardiac resynchronization therapy outcomes in patients under nonoptimal medical therapy. *J. Arrhythmia* **2018**, *34*, 548–555. [CrossRef]
37. Hori, M.; Okamoto, H. Heart rate as a target of treatment of chronic heart failure. *J. Cardiol.* **2012**, *60*, 86–90. [CrossRef]
38. Urso, C.; Brucculeri, S.; Caimi, G. Acid–base and electrolyte abnormalities in heart failure: Pathophysiology and implications. *Heart Fail. Rev.* **2015**, *20*, 493–503. [CrossRef]
39. Forleo, G.B.; Santini, L.; Campoli, M.; Malavasi, M.; Scaccia, A.; Menichelli, M.; Riva, U.; Lamberti, F.; Carreras, G.; Orazi, S.; et al. Long-term monitoring of respiratory rate in patients with heart failure: The Multiparametric Heart Failure Evaluation in Implantable Cardioverter-Defibrillator Patients (MULTITUDE-HF) study. *J. Interv. Card. Electrophysiol.* **2015**, *43*, 135–144. [CrossRef]
40. Seith, A.K. A MATLAB toolbox for Granger causal connectivity analysis. *J. Neurosci. Methods* **2010**, *186*, 262–273. [CrossRef]
41. Nemati, S.; Edwards, B.A.; Lee, J.; Pittman-Polletta, B.; Butler, J.P.; Malhotra, A. Respiration and heart rate complexity: Effects of age and gender assessed by band-limited transfer entropy. *Respir. Physiol. Neurobiol.* **2013**, *189*, 27–33. [CrossRef]
42. Iatsenko, D.; Bernjak, A.; Stankovski, T.; Shiogai, Y.; Owen-Lynch, P.J.; Clarkson, P.B.M.; McClintock, P.V.E.; Stefanovska, A. Evolution of cardiorespiratory interactions with age. *Philos. Trans. R. Soc. A Math. Phys. Eng. Sci.* **2013**, *371*, 20110622. [CrossRef]
43. O'mahony, D.; Bennett, C.; Green, A.; Sinclair, A. Reduced baroreflex sensitivity in elderly humans is not due to efferent autonomic dysfunction. *Clin. Sci.* **2000**, *98*, 103. [CrossRef]
44. Gerritsen, J.; TenVoorde, B.J.; Dekker, J.M.; Kostense, P.J.; Bouter, L.M.; Heethaar, R.M. Baroreflex sensitivity in the elderly: Influence of age, breathing and spectral methods. *Clin. Sci.* **2000**, *99*, 371–381. [CrossRef]
45. Breston, L.; Leonardis, E.J.; Quinn, L.K.; Tolston, M.; Wiles, J.; Chiba, A.A. Convergent cross sorting for estimating dynamic coupling. *Sci. Rep.* **2021**, *11*, 20374. [CrossRef] [PubMed]
46. Widjaja, D.; Montalto, A.; Vlemincx, E.; Marinazzo, D.; Van Huffel, S.; Faes, L. Cardiorespiratory Information Dynamics during Mental Arithmetic and Sustained Attention. *PLoS ONE* **2015**, *10*, e0129112. [CrossRef] [PubMed]

entropy

Brief Report

Bradycardia May Decrease Cardiorespiratory Coupling in Preterm Infants

Miguel Ángel Porta-García [1,2,†], Alberto Quiroz-Salazar [2,†], Eric Alonso Abarca-Castro [3] and José Javier Reyes-Lagos [2,*]

1 Center of Research and Innovation in Information Technology and Communication—INFOTEC, Mexico City 14050, Mexico; miguel.porta@infotec.mx
2 School of Medicine, Autonomous University of the State of Mexico (UAEMéx), Toluca de Lerdo 50180, Mexico; aquirozs002@alumno.uaemex.mx
3 Department of Health Sciences, Metropolitan Autonomous University-Lerma (UAM-L), Lerma de Villada 52005, Mexico; e.abarca@correo.ler.uam.mx
* Correspondence: jjreyesl@uaemex.mx
† These authors contributed equally to this work.

Abstract: Bradycardia, frequently observed in preterm infants, presents significant risks due to the immaturity of their autonomic nervous system (ANS) and respiratory systems. These infants may face cardiorespiratory events, leading to severe complications like hypoxemia and neurodevelopmental disorders. Although neonatal care has advanced, the influence of bradycardia on cardiorespiratory coupling (CRC) remains elusive. This exploratory study delves into CRC in preterm infants, emphasizing disparities between events with and without bradycardia. Using the Preterm Infant Cardio-Respiratory Signals (PICS) database, we analyzed interbeat (R-R) and inter-breath intervals (IBI) from 10 preterm infants. The time series were segmented into bradycardic (B) and non-bradycardic (NB) segments. Employing information theory measures, we quantified the irregularity of cardiac and respiratory time series. Notably, B segments had significantly lower entropy values for R-R and IBI than NB segments, while mutual information was higher in NB segments. This could imply a reduction in the complexity of respiratory and cardiac dynamics during bradycardic events, potentially indicating weaker CRC. Building on these insights, this research highlights the distinctive physiological characteristics of preterm infants and underscores the potential of emerging non-invasive diagnostic tools.

Keywords: cardiorespiratory coupling; bradycardia; apnea of prematurity; neonatal care; electrocardiogram; respiratory signal; neurodevelopment; information theory

1. Introduction

Each year, approximately 15 million babies are born prematurely before 37 weeks of gestation. These infants are susceptible to various health issues, which are the leading cause of death in children under the age of 5 [1].

The function of the cardiac and respiratory systems is influenced and regulated by the autonomic nervous system (ANS). The ongoing activity of the ANS can result in interactions between the cardiac and respiratory rhythms, also known as cardiorespiratory interactions (CRIs). These CRIs can lead to relative synchronization between the two systems, called cardiorespiratory coupling (CRC). It is acknowledged that the respiratory rhythm takes precedence over the cardiac rhythm [2]. CRC, which denotes the interplay between the pulmonary and cardiovascular systems, is mainly correlated with efferent vagal activity and is an intrinsic process that can have advantageous impacts on the general well-being of living beings by establishing collaborative associations between the cardiovascular and respiratory systems. However, there is insufficient knowledge regarding the mechanisms that promote normal and abnormal interactions between the two systems, potentially resulting in ANS alterations in various diseases [3]. Notably, CRC decreases when there is a shift in the balance between sympathetic and vagal activity, characterized by an increase in sympathetic activity and a decrease in vagal activity, as indicated in relevant research [4]. This observation persists even in individuals who have undergone training [5].

Bradycardia, as defined by the Neonatal Resuscitation Program (NRP), refers to a neonate's heart rate dropping below 100 beats per minute (bpm) and commonly occurs in preterm infants [6]. Among all the potential complications associated with prematurity, those related to CRC may include apnea. Apnea is typically characterized by a cessation of breathing lasting at least 20 s or a shorter pause in breathing accompanied by bradycardia (a heart rate below 100 beats per minute), pallor, or cyanosis [7]. However, premature babies often experience brief apnea episodes lasting less than 20 s due to even minor interruptions in airflow, leading to hypoxemia or bradycardia.

Preterm infants face a complex array of changes, and the issue of bradycardia within the context of CRC introduces significant concern. In this way, infant bradycardia can lead to respiratory events like apnea. These events, characterized by breathing interruptions and associated bradycardia, can result in hypoxemia and cerebral hypoperfusion, posing immediate threats to the infant's well-being [8]. Furthermore, the long-term consequences extend to potential neurodevelopmental disorders, which are well documented in preterm infants [8,9]. Bradycardia contributes substantially to morbidity and mortality in this vulnerable population [10]. Understanding the intricate interplay between bradycardia and CRC is crucial for identifying potential risks to both short-term and long-term neurodevelopmental outcomes [10–12]. Authors have associated these factors with delays in neurodevelopment and numerous other factors related to prematurity associated with long-term negative consequences [13]. The incidence of bradycardia in preterm infants differs based on their gestational age, birth weight, and other risk factors, including respiratory distress syndrome, sepsis, and anemia [14,15]. Furthermore, it has been reported that there is a correlation between bradycardia, low birth weight, and gestational age [16]. Despite the progress in neonatal care, bradycardia remains a profound contributor to morbidity and mortality in premature infants [17].

Premature neonates face an increased risk of developing bradycardia due to their immature respiratory and ANS. This condition can lead to apnea episodes and periodic breathing, indicative of an underdeveloped cardiorespiratory control system [18,19]. However, some newborns born prematurely or at full term may experience a drop in heart rate during feeding, even without other signs of respiratory or gastroesophageal reflux disease. The specific role of the ANS in triggering these symptoms remains unclear. It is plausible that the decrease in heart rate during feeding is linked to increased reflex parasympathetic autonomic nervous system activity [20].

Dick et al. highlighted the interrelationship between the cardiovascular and respiratory systems, which are distinct but interconnected [21]. In recent years, integrative physiology has gained increasing attention [22], providing a broader comprehension of how physiological systems interact. Several authors have examined the importance of CRC in neonates. Joshi et al. investigated alterations in CRC among preterm infants and revealed a synchronization of heart rate acceleration and deceleration with inspiration and expiration. They emphasized the sensitivity of these changes to whether the infants were provided with kangaroo care [23]. Groot et al. studied cardiorespiratory activity and its correlation with sleep quality and duration. Their findings suggest that heart rate and respiratory frequency are linked to active and quiet sleep, while term neonates exhibit stability and stillness [24]. Similar research conducted by Lucchini et al. employed cross-spectral analysis and bivariate phase-rectified signal averaging (BPRSA) to measure the occurrence and intensity of CRC. The results of the analysis indicate that the BPRSA curve could be responsive to CRC, evaluating the lag between respiratory activity and R-R series [25].

From the literature thus far, there is an evident gap concerning the impact of bradycardia on CRC in preterm infants. Evaluating CRC is especially crucial for those infants who experience bradycardia due to an underdeveloped ANS. This study aims to assess

CRC in preterm infants utilizing nonlinear methodologies grounded in information theory, including mutual information (MI) and entropy. The objective is to explore some autonomic mechanisms associated with nonlinear interactions between the cardiovascular and respiratory systems. This evaluation was performed during bradycardia events and periods devoid of bradycardia. The central hypothesis is that CRC exhibits variances between bradycardia and non-bradycardia events in preterm infants.

2. Materials and Methods

This exploratory study used electrocardiograms (ECGs) and respiratory signals from the Preterm Infant Cardio-Respiratory Signals (PICS) database [17]. This database contains simultaneous electrocardiograms and respiratory signals from 10 preterm infants. These signals have an approximate duration ranging from 20 to 70 h. The infants had postconceptional ages (PCA) between $29\frac{3}{7}$ and $34\frac{2}{7}$ weeks and weights ranging from 843 to 2100 g (Table 1). The data were collected in the Neonatal Intensive Care Unit (NICU) at the University of Massachusetts Memorial Healthcare and are available at http://physionet.org/content/picsdb/1.0.0/ (accessed on 1 August 2023).

Table 1. PCA, weight, and mean heart rate of PICS subjects.

Subject	**1**	**2**	**3**	**4**	**5**	**6**	**7**	**8**	**9**	**10**
PCA	$29\frac{3}{7}$	$30\frac{5}{7}$	$30\frac{5}{7}$	$30\frac{1}{7}$	$32\frac{2}{7}$	$30\frac{1}{7}$	$30\frac{1}{7}$	$32\frac{3}{7}$	$30\frac{4}{7}$	$34\frac{2}{7}$
Weight (kg)	1.2	1.76	1.71	0.84	1.67	1.14	1.11	2.1	1.23	1.9
Mean heart rate (bpm)	155	131	131	167	143	137	162	141	150	156

This table is an extract from [10].

2.1. Data Processing Pipeline

The complete processing of the ECG and respiratory signals is summarized in Figure 1 and was executed using MATLAB® vR2022b (The Mathworks, Inc., Natick, MA, USA). Data extraction of the ten infants (Figure 1a) was carried out from both ECG and respiration records (provided in the standard WFDB format) using the WFDB Toolbox for MATLAB® [26,27]. The interbeat intervals (R-R) and inter-breath intervals (IBI) time series were computed based on the annotation files accompanying the PICS database for the ECG and respiratory signals, respectively (Figure 1b).

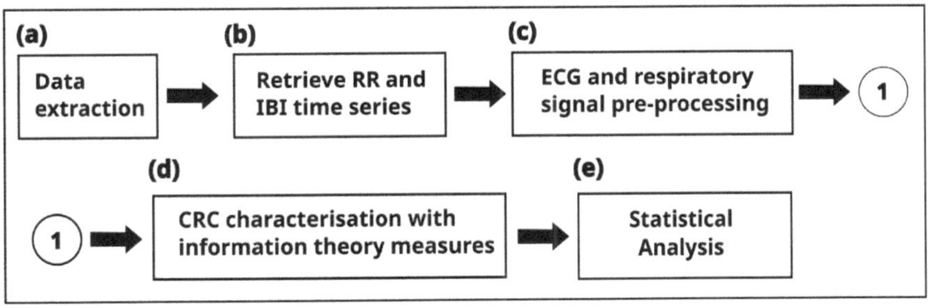

Figure 1. Cardiorespiratory signal processing pipeline.

Signal preprocessing (Figure 1c) involved several considerations. An adaptive filter was applied to the R-R and IBI time series [28] to correct for any outliers (artifacts). Since the sampling is not uniform, both time series were resampled at 4 Hz (in addition to removing high-frequency noise) to redistribute the non-equidistant samples of the R-R intervals and IBI; then, both time series were synchronous. Later, both time series were divided according to bradycardia periods. We define bradycardia as events in which the heart rate slows to less than 100 bpm (or equivalently R-R > 0.6 s) and lasts for at least two beats (>1.2 s). We categorized the data into bradycardic (B) and non-bradycardic (NB) segments based on the resampled R-R and IBI time series. We applied Inequality 1 to evaluate each reliable R-R sample:

$$(i \geq 0.6\ s)\ \&\&\ (i + (i+1)) \geq 1.2\ s \tag{1}$$

where i is the current R-R sample, the number of B data events is significantly less than the number of NB events, resulting in B events having much fewer time samples. Random Undersampling (RUS) was then used to balance the length of B and NB samples (for both R-R and IBI time series), extracting the necessary number of time samples from R-R and IBI during NB to equal the number of samples of R-R and IBI for the B condition. Remarkably, each time the adaptive filtering process is performed for each subject, the output has slightly different data because the algorithm of the adaptive filter [28] uses random values from $[\mu_a(n) - 0.5\sigma_a(n), \mu_a(n) + 0.5\sigma_a(n)]$ to replace those recognized as outliers, where μ_a is the adaptive mean and σ_a the adaptive standard deviation. This adaptive filter comprises three primary steps: (i) the identification and removal of evident recognition errors, (ii) application of the adaptive percent filter to the physiological signals, and (iii) implementation of the adaptive controlling filter. Recognition errors, such as zero-length R-R or IBI intervals and pauses, are consistently excluded from consideration based on established practices in clinical acquisition systems.

Therefore, steps from 1b to 1d were performed over 100 trials, and the input for the statistical analysis (Figure 1e) is the median. Figure 2 shows the samples extracted from the R-R and IBI time series, respectively. Since information theory measures are related to probabilities and distributions, the original data are organized into bins (similar to a histogram) and transformed into probability values, as shown in the data distribution plot in Figure 2.

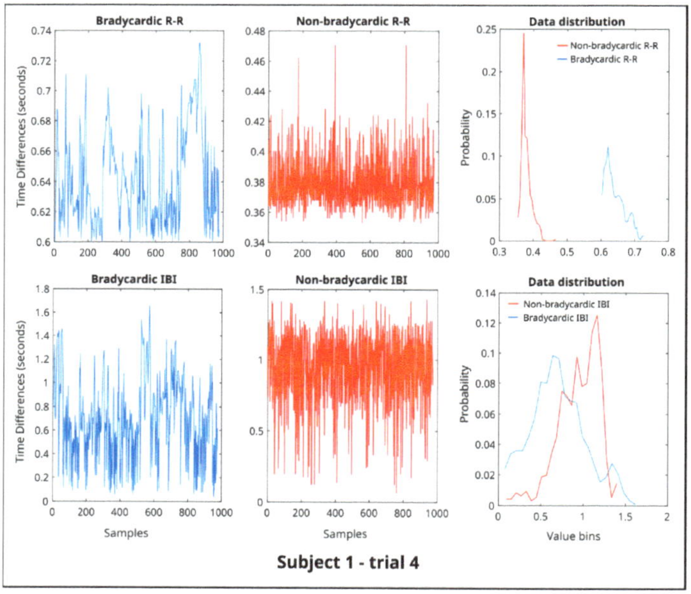

Figure 2. Example of the data distribution for both interbeat intervals (R-R) and inter-breath intervals (IBI) time series during bradycardic (B) and non-bradycardic (NB) conditions (Subject 1, trial 4).

2.2. *Information Theory Measures*

2.2.1. Notation and Preliminaries

Information theory metrics formally define dynamics and possible interactions between systems X and Y (for bivariate cases), modeled as discrete random variables. Then, X and Y can take values from their respective alphabets A_x and A_y according to a probability distribution denoted by $p(x) = PrPr\ \{X = x\}\ ,\ x \in A_x$ and $p(y) = PrPr\ \{Y = y\}\ ,\ y \in A_y$. In [29], Shannon developed a theory based on a statistical representation of a communication system. Shannon extended the information measure introduced by Ralph Hartley and linked this concept to a probability distribution associated with the source of symbols. According to Shannon, the information I associated with observing a symbol with a probability of occurrence p is represented by $I(p) = log\frac{1}{p}$. Thus, the information obtained from observing a symbol depends on its probability of occurrence. In other words, the less likely the symbol is to occur, the more information is gained from observing it.

The key measure in information theory is the Shannon entropy. It quantifies the information content of a random variable X (it can be extended to a vector of two or more random variables) as the average ambiguity associated with its outcomes [29]. Entropy measures were calculated to characterize the irregularity of the R-R and IBI time series within both B and NB segments. Additionally, mutual information and cross-entropy were employed to assess the degree of coupling between the R-R and IBI time series across these segments. A brief description of the information above indices is presented below. For improved clarity in the subsequent sections, we will consider X as the R-R time series and Y as the IBI time series.

2.2.2. Entropy

The entropy (referred to as Shannon entropy) $H(X)$ of a discrete random variable X with the alphabet A_x and a probability mass function $p(x) = PrPr\ \{X = x\}\ ,\ x \in A_x$ is a measure of uncertainty of occurrence of a certain event, quantifying the amount of information a variable has:

$$H(X) = -\sum_{i=1}^{n} p(x_i)log_2 p(x_i) \tag{2}$$

where H is the measure of entropy, p is the probability of observing the ith value of the bin series data x, and n is the number of bins. We consider the log in base two as bits as units of entropy.

2.2.3. Cross-Entropy

Also proposed by Kullback and called directed divergence [30], the cross-entropy $cH(X,Y)$ measures the degree of difference between two discrete random variables X and Y, with the alphabet A_x and A_y, and a probability distribution $p(x)$ and $p(y)$, respectively:

$$cH(X,Y) = -\sum_{i=1}^{n} p(x_i)log_2 p(y_i) \tag{3}$$

$$cH(Y,X) = -\sum_{i=1}^{n} p(y_i)log_2 p(x_i) \tag{4}$$

While there are numerous approaches to computing cross-entropy [29], we chose to use the widely accepted definition in Machine Learning [31] to avoid entering the debate over the optimal cross-entropy measure, which is beyond the scope of this paper.

2.2.4. Mutual Information

Mutual information (*MI*) is defined as a measure of the amount of information shared between two discrete random variables X and Y, with the alphabet A_x and A_y, and a joint probability distribution $p(x,y)$. It quantifies the nonlinear dependence of two signals. In terms of entropy, the MI can be calculated as the sum of two variable entropies minus the joint entropy [32]. Equation (5) shows the general form of MI.

$$MI(X,Y) = H(X) + H(Y) - H(X,Y) \tag{5}$$

where $H(X,Y)$ represents joint entropy of time series X and Y, respectively.

2.3. Statistical Analysis

The 100 trials performed on each of the ten subjects yielded 1000 values of the entropy of R-R and IBI, cross-entropy, and mutual information (*MI*) under both B and NB events. To simplify the analysis, we computed the median of these 100 measurements for each participant in both groups (B and NB), resulting in a single value per subject in each group ($n = 10$). Given the small sample size and the uncertain distribution of the outcomes, which cannot be assumed to be approximately normal, nonparametric tests were considered appropriate [33]. Therefore, a Wilcoxon matched-pairs signed rank test for repeated measures was chosen as the statistical method to assess differences. These results were gathered and processed for within-subjects analysis using GraphPad Prism 8 (GraphPad Software, La Jolla, CA, USA) software.

3. Results

Figure 3 presents the boxplots representing the average values of the information-theory indices for both bradycardia (B) and non-bradycardia (NB) samples. Initially, it was essential to ensure that each set of extracted data demonstrated the presence of either bradycardia or non-bradycardia, which was established through the distribution of the R-R intervals. An example is depicted in Figure 2.

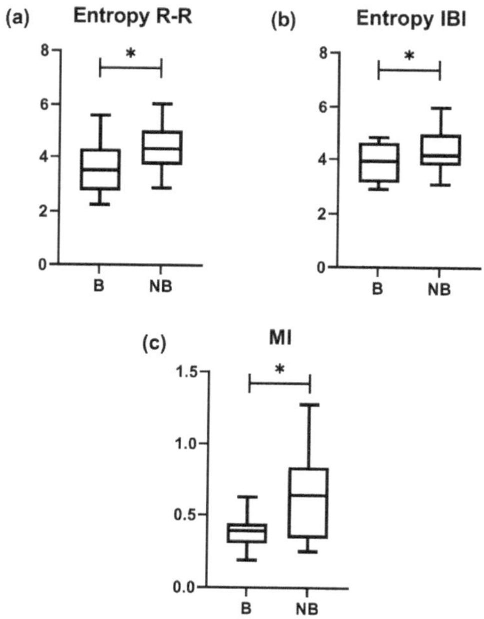

Figure 3. Boxplots of significant information-theory indices related to complexity between interbeat intervals (R-R) and inter-breath intervals (IBI) segments of preterm infants during bradycardia (B) and non-bradycardia events (NB). (**a**) Entropy of R-R time series between B and NB; (**b**) Entropy of IBI time series between B and NB; and (**c**) Mutual information (MI) of R-R and IBI between B and NB. * $p < 0.05$ between B and NB by Wilcoxon matched-pairs signed rank test.

The entropy of R-R and IBI time series, *MI*, and cross-entropy, revealed significant differences between the B and NB segments. These observed differences provide evidence, indicating decreased MI and entropy values in the B segments compared to the NB segments.

Focusing on the entropy of R-R, as illustrated in Figure 3a, the B segments were noted to have significantly lower mean values than the NB segments (3.656 ± 1.052

vs. 4.343 ± 0.9411, respectively, $p = 0.0371$). In a similar vein, the analysis of the entropy of IBI displayed in Figure 3b showed a significant decrease in mean values for the B segments in comparison to the NB segments (3.950 ± 0.7259 vs. 4.358 ± 0.8800, respectively, $p = 0.0244$).

Upon examining the mutual information indices (Figure 3c), a significant difference was identified between the B and NB segments. The B segments exhibited an average MI value of 0.3895 ± 0.1184, while the NB segments exhibited a higher average value of 0.6582 ± 0.3110. This significant difference, with a p-value of 0.0137, is visually represented in Figure 2c.

In contrast, the cross-entropy (R-R -> IBI and IBI -> R-R) did not reveal significant differences between the B and NB segments. The calculated average cross-entropy (R-R -> IBI) for B and NB was 5.804 ± 2.944 and 5.640 ± 2.531, respectively, resulting in a p-value of 0.3477. Similarly, the cross-entropy (IBI -> R-R) for B and NB was 6.701 ± 3.923 and 6.272 ± 2.766, respectively, resulting in a p-value of 0.500.

4. Discussion

While most of the research concerning bradycardia in preterm infants has been centered on predicting bradycardic/apneic events before their occurrence [17,34–36], our study stands out as the exploratory effort in assessing CRC in preterm infants during events, whether they involve bradycardia or not. Consequently, the significance of this preliminary study lies in its contribution to advancing our comprehension of the intricate physiological cardiorespiratory mechanisms through information theory tools. Moreover, it helps us recognize potential alterations in their complex dynamics and nonlinear correlations, offering valuable insights into this crucial aspect of infant healthcare. This exploratory study was designed to elucidate further whether a notable difference exists in the interaction between the cardiovascular and respiratory systems among preterm infants during episodes with and without bradycardia. The rationale behind this investigation stems from the understanding that premature birth can result in underdeveloped coordination between these systems, with bradycardia being a potential indicator of immature control over cardiovascular and respiratory functions [37,38].

Our study's initial results indicate that the interplay between cardiac and respiratory systems may be altered during bradycardia episodes compared to non-bradycardic periods. Complexity measures applied to the R-R and IBI time series demonstrate a significative decrease during bradycardic episodes, as illustrated in Figure 3a,b, corresponding to reduced entropy. This observation is in line with prior studies that have examined abnormal neonatal cardiac conditions [39,40]. The noted reduction in complexity supports the hypothesis that pathological states, such as congestive heart failure, are characterized by less complexity in their physiological patterns than healthy systems [41]. Considering these findings, the link between bradycardia and a decreased complexity in the cardiac and respiratory systems—as suggested by the lower entropy values in R-R and IBI segments during bradycardic periods—warrants careful interpretation. Our study does not definitively establish causality. Furthermore, the lack of significant difference in cross-entropy regarding the influence of respiration on R-R intervals (or the inverse) highlights the need for a more detailed investigation into their interconnection. Subsequent research is essential to elucidate the underlying mechanisms and to potentially confirm a causal relationship between the complexity of the cardiac and respiratory systems and the incidence of bradycardia in preterm infants.

Mutual information, derived from information theory, measures the potential nonlinear interactions between two variables—here, the R-R intervals of the heart and the IBI from respiration. The *MI* values observed in the NB segments could suggest nonlinear associations between the cardiac and respiratory systems when bradycardia is absent. Such a relationship might imply a level of synchronization between the heart and lungs that could be important for physiological stability. Conversely, the *MI* values in the B (bradycardic) segments may reflect a different degree of interaction during bradycardic episodes, which the nature of bradycardia itself might influence. The slower heart rate associated with bradycardia could potentially affect the usual synchrony between the cardiovascular and respiratory systems, possibly leading to a change in their capacity to adjust and communicate effectively with one another, a process that is crucial for maintaining balance within the body and adapting to changes in the environment [3]. However, these interpretations remain speculative, and the results do not definitively establish a difference in cardiorespiratory coupling (CRC) between bradycardic and non-bradycardic states. Further investigation is needed to understand these associations fully.

In light of these speculative interpretations, it is well recognized that the heart and lungs are engaged in a constant reciprocal interaction, forming a functional and anatomical reserve that underpins CRC. This dynamic exchange is key to the seamless coordination between these systems under typical conditions. Disruptions in this coupling may manifest in the face of various cardiac or pulmonary disorders, including bradycardia. Therefore, the findings of our study, which hint at altered interactions during bradycardic episodes, could indicate such disruptions. This notion aligns with the established concept of ongoing interaction between the heart and lungs, yet the exact relationship with bradycardia observed here remains hypothetical, underscoring the need for further research to clarify the nature of this linkage.

These findings underscore the importance of comprehending in detail the dynamics of the interaction between the heart and lungs in premature infants, as this understanding can have significant implications for the health and neurodevelopment of this population. Identifying and addressing alterations in this interaction could help prevent or mitigate issues and delays in the neurodevelopment of premature infants [12]. Furthermore, this research emphasizes the need to consider complex and nonlinear factors when assessing the health of premature infants, as these factors can substantially impact their short-term and long-term well-being.

Building on this, it is widely acknowledged in existing research that preterm infants are at a higher risk for various health issues and suboptimal neurodevelopmental outcomes, the mechanisms of which remain incompletely understood. A related study evaluated the autonomic development of newborns based on their gestational age at birth, with a specific focus on regulating cardiorespiratory functions. The research suggests that infants born prematurely at 35–36 weeks of gestational age exhibit a higher susceptibility to breathing instability, evidenced by diminished CRC [42]. It underscores the observation that an increase in gestational age at birth corresponds to an enhanced coupling between the two systems. These findings highlight the lower level of maturation in preterm infants and their underdeveloped cardiorespiratory regulation, which aligns with epidemiological data and may indicate a heightened risk for divergent outcomes [42]. Given the exploratory nature of our research, it suggests a direction for future studies to investigate further and possibly expand upon the understanding of cardiorespiratory interactions in preterm infants, with a particular focus on bradycardia. A study utilizing the same database for predicting bradycardic events revealed that increased variance in the heart rate signal was a precursor to severe bradycardia. This increased variance was associated with a rise in power from low content dynamics in the low-frequency and diminished multiscale entropy values preceding bradycardia [17]. This observation amplifies our findings, pointing towards nuanced physiological shifts before the onset of bradycardic episodes. Notably, previous findings have shown a significant negative correlation between entropy and the duration of respiratory support required, further emphasizing the potential clinical implications of these observations [43].

Furthermore, additional research has explored the ramifications of initial immunization on cardiorespiratory events in extremely preterm infants. The observed cardiorespiratory events were mainly attributed to a dominant sympathetic influence over heart rate, reduced Heart Rate Variability (HRV), characterized by low entropy, and a continuing immaturity in controlling respiratory rhythm [44]. These findings reiterate the complexity of the regulatory mechanisms in preterm infants and their susceptibility to alterations in physiological states. Another significant contribution to this discourse was a study highlighting the predictive capabilities of cardiorespiratory variability parameters during non-invasive respiratory support in extremely preterm infants. The study demonstrated moderate predictive accuracy for successful extubation, with these measures showcasing notable differences between successful and unsuccessful extubation cases. This highlights their potential as biomarkers to assess extubation outcomes in this vulnerable group [45].

The uniqueness of our work lies in its application of information theory tools to assess alterations in CRC from recorded data, which may reveal patterns that are not immediately apparent in predictive models. This retrospective analysis could inform the development of sophisticated monitoring techniques that, while not real-time, could still enhance clinicians' ability to understand and respond to bradycardic episodes or other cardiorespiratory anomalies. Additionally, by retrospectively assessing CRC during bradycardia, our study offers fresh insights into the physiological states of preterm infants. We deepen our understanding of the cardiorespiratory system's function during these critical periods by examining the reduced complexity and potential nonlinear correlations between cardiac and respiratory signals. This knowledge could lead to improved clinical strategies to enhance this vulnerable population's health and developmental outcomes.

The current study represents a novel approach in understanding the interplay between bradycardia and CRC in preterm infants. Previous studies in this field, such as those by Faes et al., 2014 [46], Rozo et al., 2021 [47], Lucchini et al., 2020 [48], and Lucchini et al., 2018 [25], have laid important groundwork. Faes et al.'s study focused on the use of model-free tools for time series analysis to understand physiological system interactions, introducing a method to evaluate the direction, magnitude, and timing of information transfer between systems. Rozo et al. compared various methods to estimate Transfer Entropy (TE) in cardio-respiratory interactions, finding adaptive partitioning most effective. Their work emphasized the importance of choosing appropriate signals and methods for analyzing such interactions. The study by Lucchini et al., 2020, explored cardiorespiratory information transfer in healthy neonates, aiming to describe its development relative to gestational age. They extended the traditional TE measure to analyze both instantaneous and delayed effects between cardiac and respiratory systems. Lastly, Lucchini et al., 2018, investigated the phase coupling and its directionality in newborn infants, assessing the influence of gestational age at birth on the development of this synchronization.

The current study builds upon these foundations by specifically focusing on the differences in CRC during bradycardic and non-bradycardic events in very preterm infants. Unlike the previous studies which primarily involved healthy subjects or general neonate populations, this research delves into a more vulnerable group, very preterm infants experiencing bradycardia. The use of the PICS database for analyzing cardiac and respiratory time series, coupled with information theory measures, marks a distinctive approach. This study's findings about the lower entropy values in bradycardic segments and the implications for reduced complexity in cardiorespiratory dynamics during such events contribute significantly to the understanding of autonomic maturation and the interplay between cardiac and respiratory systems in this high-risk population. This direction is crucial for developing better diagnostic tools and enhancing healthcare outcomes for preterm infants.

CRC assessment may help provide additional measures of infant well-being, which would be complementary in a clinical setting. To follow up these investigations in the future, CRC features characterized by information theory measures could be used to implement classifiers to distinguish bradycardic periods from regular ones. Although at first glance, it may seem redundant to develop classifiers to distinguish bradycardia, as it can be easily identified through heart rate, there are merits in considering this approach. Bradycardia can be readily detected, but identifying subtle patterns, correlations, and nonlinear features that could indicate the severity, etiology, or prognostic implications of bradycardia may not be as straightforward. The classifiers developed from information theory measures could identify these features and provide a deeper understanding and earlier detection of cardiorespiratory irregularities. This could justify the effort and resources needed to develop and implement such tools in a clinical setting.

Despite the limitations inherent in using electrocardiograms and respiratory signals, these remain standard and widely accepted methods for assessing CRC [49–51]. Advances in technology have improved the accuracy of these measurements, and the use of signal-processing techniques can mitigate the impact of artifacts. Furthermore, complementary methods and additional physiological measures can be incorporated to validate findings and ensure the robustness of the results.

Finally, these novel measures might become non-invasive complementary diagnostic tools to investigate the physiological mechanisms involved in bradycardia and potential predictors of bradycardic periods. Future work consists of the computation of relevant indexes such as functional autonomic age (FAA) derived from the ECG signal [42] and respiratory signs; it offers direct avenues towards estimating autonomic maturation at the bedside during intensive care monitoring.

5. Conclusions

This preliminary study provides an exploratory look at cardiorespiratory coupling in preterm infants, comparing episodes with and without bradycardia. The changes noted in the entropy of interbeat (R-R) and inter-breath (IBI) interval time series might reflect a different organization in the respiratory and cardiac dynamics during bradycardic events. Moreover, the observed variations in mutual information may suggest potential changes in the cardiorespiratory interaction during such events. Our results could imply a reduction in the complexity of respiratory and cardiac dynamics during bradycardic events, potentially indicating weaker CRC. These initial observations contribute to a nuanced understanding of the physiological intricacies of preterm infants and highlight the possible utility of new metrics as non-invasive diagnostic tools. The noted variations in entropy raise questions about possible neurodevelopmental concerns in preterm infants. These preliminary findings emphasize the importance of further research to explore the broader implications of cardiorespiratory interactions on these infants' neurodevelopment and overall health. The study suggests directions for future research that may advance our understanding of autonomic growth and cardiorespiratory relationships, ultimately aiming to improve healthcare outcomes for preterm infants.

Author Contributions: Conceptualization, J.J.R.-L.; methodology, M.Á.P.-G. and A.Q.-S.; software, M.Á.P.-G. and A.Q.-S.; validation, M.Á.P.-G. and A.Q.-S.; formal analysis, J.J.R.-L., M.Á.P.-G. and A.Q.-S.; investigation, A.Q.-S.; resources, J.J.R.-L.; data curation, A.Q.-S.; writing—original draft preparation, J.J.R.-L., M.Á.P.-G. and A.Q.-S.; writing—review and editing, J.J.R.-L., M.Á.P.-G., A.Q.-S. and E.A.A.-C.; visualization, M.Á.P.-G. and A.Q.-S.; supervision, J.J.R.-L. and M Á.P.-G.; project administration, M.Á.P.-G.; funding acquisition, J.J.R.-L. All authors have read and agreed to the published version of the manuscript.

Funding: This research was funded by INFOTEC.

Institutional Review Board Statement: Not applicable.

Data Availability Statement: Data are contained within the article.

Acknowledgments: Financial support from the Mexico State Council of Science and Technology (COMECYT) for M.Á.P.-G. is gratefully acknowledged (fellowship CAT2022-0032).

Conflicts of Interest: The authors declare no conflict of interest.

References

1. WHO. Children: Improving Survival and Well-Being. Available online: https://www.who.int/news-room/fact-sheets/detail/children-reducing-mortality (accessed on 1 September 2023).
2. Sobiech, T.; Buchner, T.; Krzesiński, P.; Gielerak, G. Cardiorespiratory Coupling in Young Healthy Subjects. *Physiol. Meas.* **2017**, *38*, 2186–2202. [CrossRef] [PubMed]

3. Garcia, A.J.; Koschnitzky, J.E.; Dashevskiy, T.; Ramirez, J.-M. Cardiorespiratory Coupling in Health and Disease. *Auton. Neurosci.* **2013**, *175*, 26–37. [CrossRef] [PubMed]
4. Porta, A.; Bassani, T.; Bari, V.; Tobaldini, E.; Takahashi, A.C.; Catai, A.M.; Montano, N. Model-Based Assessment of Baroreflex and Cardiopulmonary Couplings during Graded Head-up Tilt. *Comput. Biol. Med.* **2012**, *42*, 298–305. [CrossRef] [PubMed]
5. de Abreu, R.M.; Catai, A.M.; Cairo, B.; Rehder-Santos, P.; da Silva, C.D.; Signini, D.F.; Sakaguchi, C.A.; Porta, A. A Transfer Entropy Approach for the Assessment of the Impact of Inspiratory Muscle Training on the Cardiorespiratory Coupling of Amateur Cyclists. *Front. Physiol.* **2020**, *11*, 134. [CrossRef] [PubMed]
6. Hasenstab-Kenney, K.A.; Sanchez, J.B.; Prabhakar, V.; Lang, I.M.; Shaker, R.; Jadcherla, S.R. Mechanisms of Bradycardia in Premature Infants: Aerodigestive–Cardiac Regulatory–Rhythm Interactions. *Physiol. Rep.* **2020**, *8*, e14495. [CrossRef]
7. Eichenwald, E.C.; Watterberg, K.L.; Aucott, S.; Benitz, W.E.; Cummings, J.J.; Goldsmith, J.; Poindexter, B.B.; Puopolo, K.; Stewart, D.L.; Wang, K.S.; et al. Apnea of Prematurity. *Pediatrics* **2016**, *137*, e20153757. [CrossRef] [PubMed]
8. Pichler, G.; Urlesberger, B.; Ller, W.M. Impact of Bradycardia on Cerebral Oxygenation and Cerebral Blood Volume during Apnoea in Preterm Infants. *Physiol. Meas.* **2003**, *24*, 671–680. [CrossRef]
9. Yee, A.K.; Siriwardhana, L.S.; Nixson, G.M.; Walter, L.M.; Wong, F.Y.; Horne, R.S.C. Developmental Consequences of Short Apneas and Periodic Breathing in Preterm Infants. *J. Perinatol.* **2023**, *43*, 1420–1428. [CrossRef]
10. Pillekamp, F.; Hermann, C.; Keller, T.; von Gontard, A.; Kribs, A.; Roth, B. Factors Influencing Apnea and Bradycardia of Prematurity—Implications for Neurodevelopment. *Neonatology* **2007**, *91*, 155–161. [CrossRef]
11. Janvier, A.; Khairy, M.; Kokkotis, A.; Cormier, C.; Messmer, D.; Barrington, K.J. Apnea Is Associated with Neurodevelopmental Impairment in Very Low Birth Weight Infants. *J. Perinatol.* **2004**, *24*, 763–768. [CrossRef]
12. Williamson, M.; Poorun, R.; Hartley, C. Apnoea of Prematurity and Neurodevelopmental Outcomes: Current Understanding and Future Prospects for Research. *Front. Pediatr.* **2021**, *9*, 755677. [CrossRef] [PubMed]
13. Stoll, B.J.; Hansen, N.I.; Bell, E.F.; Walsh, M.C.; Carlo, W.A.; Shankaran, S.; Laptook, A.R.; Sánchez, P.J.; Van Meurs, K.P.; Wyckoff, M.; et al. Trends in Care Practices, Morbidity, and Mortality of Extremely Preterm Neonates, 1993–2012. *JAMA-J. Am. Med. Assoc.* **2015**, *314*, 1039–1051. [CrossRef] [PubMed]
14. Sadoughi, A.; Shamsollahi, M.B.; Fatemizadeh, E.; Beuchée, A.; Hernández, A.I.; Ghahjaverestan, N.M. Detection of Apnea Bradycardia from ECG Signals of Preterm Infants Using Layered Hidden Markov Model. *Ann. Biomed. Eng.* **2021**, *49*, 2159–2169. [CrossRef] [PubMed]
15. Doyen, M.; Hernández, A.I.; Flamant, C.; Defontaine, A.; Favrais, G.; Altuve, M.; Laviolle, B.; Beuchée, A.; Carrault, G.; Pladys, P. Early Bradycardia Detection and Therapeutic Interventions in Preterm Infant Monitoring. *Sci. Rep.* **2021**, *11*, 10486. [CrossRef]
16. Sbrollini, A.; Mancinelli, M.; Marcantoni, I.; Morettini, M.; Burattini, L. Bradycardia Assessment in Preterm Infants. In Proceedings of the IFMBE, Portorož, Slovenia, 29 November–3 December 2020; Volume 76.
17. Gee, A.H.; Barbieri, R.; Paydarfar, D.; Indic, P. Predicting Bradycardia in Preterm Infants Using Point Process Analysis of Heart Rate. *IEEE Trans. Biomed. Eng.* **2017**, *64*, 2300–2308. [CrossRef]
18. Zhao, L.; Yang, L.; Su, Z.; Liu, C. Cardiorespiratory Coupling Analysis Based on Entropy and Cross-Entropy in Distinguishing Different Depression Stages. *Front. Physiol.* **2019**, *10*, 359. [CrossRef]
19. Veerappan, S.; Rosen, H.; Craelius, W.; Curcie, D.; Hiatt, M.; Hegyi, T. Spectral Analysis of Heart Rate Variability in Premature Infants with Feeding Bradycardia. *Pediatr. Res.* **2000**, *47*, 659–662. [CrossRef]
20. Behrman, R.E.; Butler, A.S. Mortality and Acute Complications in Preterm Infants. In *Preterm Birth: Causes, Consequences & Prevention*; National Academy of Sciences: Bethesda, MD, USA, 2006.
21. Dick, T.E.; Morris, K.F. Quantitative Analysis of Cardiovascular Modulation in Respiratory Neural Activity. *J. Physiol.* **2004**, *556*, 959–970. [CrossRef]
22. Acampa, M.; Voss, A.; Bojić, T. Editorial: Cardiorespiratory Coupling-Novel Insights for Integrative Biomedicine. *Front. Neurosci.* **2021**, *15*, 671900. [CrossRef]
23. Joshi, R.; Kommers, D.; Long, X.; Feijs, L.; Van Huffel, S.; van Pul, C.; Andriessen, P. Cardiorespiratory Coupling in Preterm Infants. *J. Appl. Physiol.* **2019**, *126*, 202–213. [CrossRef]
24. de Groot, E.R.; Knoop, M.S.; Hoogen, A.v.D.; Wang, X.; Long, X.; Pillen, S.; Benders, M.; Dudink, J. The Value of Cardiorespiratory Parameters for Sleep STATE classification in Preterm Infants: A Systematic Review. *Sleep Med. Rev.* **2021**, *58*, 101462. [CrossRef] [PubMed]
25. Lucchini, M.; Pini, N.; Fifer, W.P.; Burtchen, N.; Signorini, M.G. Characterization of Cardiorespiratory Phase Synchronization and Directionality in Late Premature and Full Term Infants. *Physiol. Meas.* **2018**, *39*, 064001. [CrossRef] [PubMed]
26. Silva, I.; Moody, G.B. An Open-Source Toolbox for Analysing and Processing PhysioNet Databases in MATLAB and Octave. *J. Open Res. Softw.* **2014**, *2*, e27. [CrossRef] [PubMed]
27. Goldberger, A.L.; Amaral, L.A.N.; Glass, L.; Hausdorff, J.M.; Ivanov, P.C.; Mark, R.G.; Mietus, J.E.; Moody, G.B.; Peng, C.-K.; Stanley, H.E. PhysioBank, PhysioToolkit, and PhysioNet: Components of a New Research Resource for Complex Physiologic Signals. *Circulation* **2000**, *101*, E215–E220. [CrossRef] [PubMed]
28. Wessel, N.; Voss, A.; Malberg, H.; Ziehmann, C.; Voss, H.U.; Schirdewan, A.; Meyerfeldt, U.; Kurths, J. Nonlinear Analysis of Complex Phenomena in Cardiological Data. *Herzschrittmachertherapie + Elektrophysiologie* **2000**, *11*, 159–173. [CrossRef]
29. Faes, L.; Porta, A.; Nollo, G. Information Decomposition in Bivariate Systems: Theory and Application to Cardiorespiratory Dynamics. *Entropy* **2015**, *17*, 277–303. [CrossRef]

30. Mallows, C.L.; Kullback, S. Information Theory and Statistics. *J. R. Stat. Soc. Ser. A (General)* **1959**, *122*, 380. [CrossRef]
31. Cui, Y.; Zhai, J.; Wang, X. Extreme Learning Machine Based on Cross Entropy. In Proceedings of the International Conference on Machine Learning and Cybernetics, Jeju Island, Republic of Korea, 19–24 June 2016; Volume 2.
32. Schulz, S.; Adochiei, F.-C.; Edu, I.-R.; Schroeder, R.; Costin, H.; Bär, K.-J.; Voss, A. Cardiovascular and Cardiorespiratory Coupling Analyses: A Review. *Philos. Trans. R. Soc. A Math. Phys. Eng. Sci.* **2013**, *371*, 20120191. [CrossRef]
33. Furfey, P.H.; Siegel, S. Nonparametric Statistics for the Behavioral Sciences. *Am. Cathol. Sociol. Rev.* **1957**, *18*, 163. [CrossRef]
34. Jiang, H.; Salmon, B.P.; Gale, T.J.; Dargaville, P.A. Prediction of Bradycardia in Preterm Infants Using Artificial Neural Networks. *Mach. Learn. Appl.* **2022**, *10*, 100426. [CrossRef]
35. Zuzarte, I.; Sternad, D.; Paydarfar, D. Predicting Apneic Events in Preterm Infants Using Cardio-Respiratory and Movement Features. *Comput. Methods Programs Biomed.* **2021**, *209*, 106321. [CrossRef] [PubMed]
36. Lim, K.; Jiang, H.; Marshall, A.P.; Salmon, B.; Gale, T.J.; Dargaville, P.A. Predicting Apnoeic Events in Preterm Infants. *Front. Pediatr.* **2020**, *8*, 570. [CrossRef] [PubMed]
37. Di Fiore, J.M.; Poets, C.F.; Gauda, E.; Martin, R.J.; MacFarlane, P. Cardiorespiratory Events in Preterm Infants: Etiology and Monitoring Technologies. *J. Perinatol.* **2016**, *36*, 165–171. [CrossRef] [PubMed]
38. Gauda, E.B.; McLemore, G.L. Premature Birth, Homeostatic Plasticity and Respiratory Consequences of Inflammation. *Respir. Physiol. Neurobiol.* **2020**, *274*, 103337. [CrossRef] [PubMed]
39. Lake, D.E.; Richman, J.S.; Griffin, M.P.; Moorman, J.R. Sample Entropy Analysis of Neonatal Heart Rate Variability. *Am. J. Physiol.-Regul. Integr. Comp. Physiol.* **2002**, *283*, R789–R797. [CrossRef] [PubMed]
40. Pravisani, G.; Beuchee, A.; Mainardi, L.; Carrault, G. Short Term Prediction of Severe Bradycardia in Premature Newborns. In Proceedings of the Computers in Cardiology, Thessaloniki, Greece, 21–24 September 2003; Volume 30.
41. Goldberger, A.L.; Amaral, L.A.N.; Hausdorff, J.M.; Ivanov, P.C.; Peng, C.-K.; Stanley, H.E. Fractal Dynamics in Physiology: Alterations with Disease and Aging. *Proc. Natl. Acad. Sci. USA* **2002**, *99* (Suppl. S1), 2466–2472. [CrossRef]
42. Lucchini, M.; Burtchen, N.; Fifer, W.P.; Signorini, M.G. Multi-Parametric Cardiorespiratory Analysis in Late-Preterm, Early-Term, and Full-Term Infants at Birth. *Med. Biol. Eng. Comput.* **2019**, *57*, 99–106. [CrossRef]
43. Jost, K.; Datta, A.N.; Frey, U.P.; Suki, B.; Schulzke, S.M. Heart Rate Fluctuation after Birth Predicts Subsequent Cardiorespiratory Stability in Preterm Infants. *Pediatr. Res.* **2019**, *86*, 348–354. [CrossRef]
44. Mialet-Marty, T.; Beuchée, A.; Ben Jmaa, W.; N'Guyen, N.; Navarro, X.; Porée, F.; Nuyt, A.M.; Pladys, P. Possible Predictors of Cardiorespiratory Events after Immunization in Preterm Neonates. *Neonatology* **2013**, *104*, 151–155. [CrossRef]
45. Latremouille, S.; Bhuller, M.; Shalish, W.; Sant'anna, G. Cardiorespiratory Measures Shortly after Extubation and Extubation Outcomes in Extremely Preterm Infants. *Pediatr. Res.* **2023**, *93*, 1687–1693. [CrossRef]
46. Faes, L.; Marinazzo, D.; Montalto, A.; Nollo, G. Lag-Specific Transfer Entropy as a Tool to Assess Cardiovascular and Cardiorespiratory Information Transfer. *IEEE Trans. Biomed. Eng.* **2014**, *61*, 2556–2568. [CrossRef] [PubMed]
47. Rozo, A.; Morales, J.; Moeyersons, J.; Joshi, R.; Caiani, E.G.; Borzée, P.; Buyse, B.; Testelmans, D.; Van Huffel, S.; Varon, C. Benchmarking Transfer Entropy Methods for the Study of Linear and Nonlinear Cardio-Respiratory Interactions. *Entropy* **2021**, *23*, 939. [CrossRef] [PubMed]
48. Lucchini, M.; Pini, N.; Burtchen, N.; Signorini, M.G.; Fifer, W.P. Transfer Entropy Modeling of Newborn Cardiorespiratory Regulation. *Front. Physiol.* **2020**, *11*, 1095. [CrossRef] [PubMed]
49. Pompe, B.; Blidh, P.; Hoyer, D.; Eiselt, M. Using Mutual Information to Measure Coupling in the Cardiorespiratory System. *IEEE Eng. Med. Biol. Mag.* **1998**, *17*, 32–39. [CrossRef]
50. Hoyer, D.; Leder, U.; Hoyer, H.; Pompe, B.; Sommer, M.; Zwiener, U. Mutual Information and Phase Dependencies: Measures of Reduced Nonlinear Cardiorespiratory Interactions after Myocardial Infarction. *Med. Eng. Phys.* **2002**, *24*, 33–43. [CrossRef]
51. Iyer, K.K.; Leitner, U.; Giordano, V.; Roberts, J.A.; Vanhatalo, S.; Klebermass-Schrehof, K.; Stevenson, N.J. Bedside Tracking of Functional Autonomic Age in Preterm Infants. *Pediatr. Res.* **2022**, *94*, 206–212. [CrossRef]

entropy

MDPI

Review

Topological Data Analysis in Cardiovascular Signals: An Overview

Enrique Hernández-Lemus [1,2,*], Pedro Miramontes [1,3] and Mireya Martínez-García [4]

1 Computational Genomics Division, National Institute of Genomic Medicine, Mexico City 14610, Mexico; pmv@ciencias.unam.mx
2 Center for Complexity Sciences, Universidad Nacional Autónoma de México, Mexico City 04510, Mexico
3 Department of Mathematics, Sciences School, Universidad Nacional Autónoma de México, Mexico City 04510, Mexico
4 Department of Immunology, National Institute of Cardiology, Mexico City 14080, Mexico; mireya.martinez@cardiologia.org.mx
* Correspondence: ehernandez@inmegen.gob.mx

Abstract: Topological data analysis (TDA) is a recent approach for analyzing and interpreting complex data sets based on ideas a branch of mathematics called algebraic topology. TDA has proven useful to disentangle non-trivial data structures in a broad range of data analytics problems including the study of cardiovascular signals. Here, we aim to provide an overview of the application of TDA to cardiovascular signals and its potential to enhance the understanding of cardiovascular diseases and their treatment in the form of a literature or narrative review. We first introduce the concept of TDA and its key techniques, including persistent homology, Mapper, and multidimensional scaling. We then discuss the use of TDA in analyzing various cardiovascular signals, including electrocardiography, photoplethysmography, and arterial stiffness. We also discuss the potential of TDA to improve the diagnosis and prognosis of cardiovascular diseases, as well as its limitations and challenges. Finally, we outline future directions for the use of TDA in cardiovascular signal analysis and its potential impact on clinical practice. Overall, TDA shows great promise as a powerful tool for the analysis of complex cardiovascular signals and may offer significant insights into the understanding and management of cardiovascular diseases.

Keywords: topological data analysis; cardiovascular signals; alegbraic topology; persistent homology; mapper algorithm

1. Introduction: Data Analytics in Modern Cardiology

Cardiovascular diseases (CVDs) have been, for a long time, a major global health problem that has caused more deaths than any form of cancer or respiratory disease combined. The detection and prediction of CVDs is made difficult by the numerous etiological factors, complex disease pathways, and diverse clinical presentations [1,2]. However, with the advent of an enhanced capability for the generation of complex high-dimensional data from electronic medical records, mobile health devices, and imaging data, one is presented with both challenges and opportunities for data-driven discovery and research. While traditional statistical approaches for risk stratification have been broadly developed, leading to important improvements of diagnosis, prognosis and in some cases therapeutics, most of these models have limitations in terms of individualized risk prediction. Recently, data analytics, artificial intelligence and machine learning have emerged as potential solutions for overcoming the limitations of traditional approaches in the field of CVD research.

Advance analytics algorithms can have a major impact on cardiovascular disease prediction and diagnosis. CVD data, however, remain challenging for common machine learning and data analytics approaches due to the wide variety and large heterogeneity of the diverse cardiovascular signals currently being probed [3,4]. Among the several different approaches that are arising to cope with such disparate data, one that results particularly outstanding for its generality and its ability to handle data integrating diverse dynamic ranges and scales is topological data analysis [5,6].

Topological data analysis (TDA) is, in short, a family of analytic methods that has been gaining relevance and recognition to model, predict and understand the behavior of complex biomedical data. TDA is founded on the tenets of algebraic topology, a mathematical field that deals with the *shape* of data and has a set of methods for studying it [7]. In this review article, we want to present the fundamentals of TDA and its applications in the analysis of cardiovascular signals. We aim to make these techniques accessible to non-experts by explaining their theoretical foundations and surveying their use in computational cardiology. We also discuss the limitations of these methods and suggest possible ways to incorporate them into clinical care and biomedical informatics in the context of cardiovascular diseases.

Figure 1 presents a graphic overview of the main ideas, starting with one or several data sources on cardiovascular signals coming from medical devices, wearables, clinical monitors, electronic health records and so on. Data are used to generate data clouds that are turned into *Metric data sets* that are then processed and analyzed with the tools of topological data analysis (see below) to generate homology groups, persistence diagrams and signatures useful to classify signals for a deeper understanding of their phenomenology.

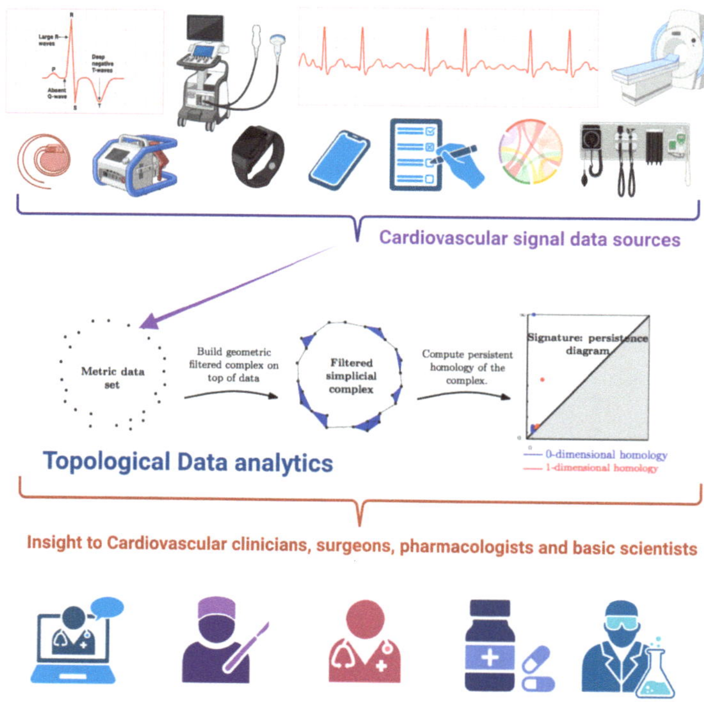

Figure 1. Topological Data Analysis in cardiology. Starting from diverse sources of cardiovascular signals, topological data analytics allowed a systematic study and categorization leading to a better understanding of the underlying phenomena, thus providing clues for clinicians and basic scientists. Figure generated with BioRender.com, central panel is taken from Larrysong, CC BY-SA 4.0 https://creativecommons.org/licenses/by-sa/4.0 (accessed on 4 June 2023), via Wikimedia Commons.

2. Fundamentals of Topological Data Analysis

Topology is a branch of mathematics that deals with the shapes of objects. It provides a framework for understanding how objects can be deformed and still retain their essential properties. For example, a circular rubber band can be stretched into an oval, but a segment of string cannot. Topologists study the connectedness of objects by counting their number of pieces and holes, and use this information to classify objects into different categories.

A related field, algebraic topology, provides a framework for describing the global structure of a space in a precise and quantitative way. It uses methods that take into account the entire space and its objects rather than just local information. Algebraic topology uses the concept of homology to classify objects based on their number and type of holes, and topological spaces, which consist of points and neighborhoods that satisfy certain conditions to do so. The notion of a topological space allows for flexibility in using topological tools in various applications, as it does not rely on numerical values to determine the proximity of points, but rather whether their neighborhoods overlap.

More formally, an *Homology* is a set of topological invariants represented by homology groups $H_k(X)$ that describe k-dimensional holes in topological space X. The rank of $H_k(X)$ (known as the kth *Betti number*), is analogous to the dimension of a vector space and indicates the number of k-dimensional holes. For example, H_0 corresponds to zero-dimensional features or connected components, H_1 corresponds to one-dimensional features or cycles, and H_2 corresponds to two-dimensional features or cavities. It is also possible to study H_k for larger values of k, but it becomes more difficult to *visualize* the associated features.

At this stage, homology seems to be a rather abstract concept; however, it can be connected in a straightforward manner to data analytics once we recognize one quite important property of data points: their *shape* as a set, that is, the way in which data points are distributed in the feature's space. One can obtain an approximation to this shape by looking at how *holes* are distributed in the data space. Our understanding of why points accumulate in one region of the data space and are missing in other regions is a powerful tool to look for trends in the data. In order to understand the topological shape of a data set and identify its holes, it is useful to assign a topological structure to the data and calculate topological invariants. Homology groups are useful for this purpose because there are efficient algorithms for computing some of the more relevant of these invariants in the context of TDA [8].

Homology groups are hence used to classify topological spaces based on the topological features of their shape, such as connectedness, loops, and voids. Homology groups of a topological space are *invariant under continuous deformations*, meaning that if two spaces have different homology groups, then they *cannot be continuously deformed into one another and are therefore topologically distinct*. Homology can thus be used to distinguish between spaces that may *appear* to be the same from other perspectives, such as those that have the same dimension or the same symmetries.

In the context of topological data analysis (TDA), the interpretation of topological features like connectedness, loops, holes, and voids involves understanding the geometric and structural properties of the data that these features represent. Let us briefly review some of these ideas.

Connectedness: Connectedness refers to the property of data points or regions being connected in a topological space. In TDA, connectedness typically corresponds to the number of connected components in a data set. The number of connected components can provide insights into the overall structure of the data. High connectedness implies that the data are relatively well-connected, while low connectedness may indicate separate clusters or isolated data points.

Loops: Loops represent closed paths or cycles in the data. They can occur when points or regions in the data form closed curves or circles. Loops can capture repetitive or periodic patterns in the data. They are often associated with cyclic structures or data points arranged in circular or ring-like formations.

Holes: Holes correspond to empty spaces or voids in the data where there are no data points. These voids can take various shapes, including spherical voids, tunnel-like voids, or irregular voids. The presence and characteristics of holes provide information about data *emptiness*. They can indicate the absence of data in specific regions or reveal patterns in data distribution, such as clustering around voids.

Voids: Voids are regions of space that lack data points. They are similar to holes but can be more generalized and may not necessarily be enclosed by data. Voids are often used to study the spatial distribution and density of data points. Large, persistent voids may suggest regions where data are scarce, while small, transient voids may highlight local fluctuations.

To interpret these topological features effectively, TDA often employs persistence diagrams or barcode diagrams. These diagrams summarize the *births* and *deaths* of topological features across a range of spatial scales, providing a quantitative way to assess the significance and persistence of these features. Here is how persistence diagrams relate to the interpretation of topological features:

Connectedness: The number of connected components is quantified by points in the persistence diagram. Longer persistence (vertical distance from birth to death) indicates more robust connected components.

Loops: Loops are associated with features in the persistence diagram. Longer loops correspond to more persistent cyclic patterns in the data.

Holes and Voids: Holes and voids are represented by clusters of points in the persistence diagram. The position of points in the diagram indicates the spatial scale and persistence of these features.

In summary, interpreting topological features in TDA involves understanding the presence, size, and persistence of connectedness, loops, holes, and voids in user data. Persistence diagrams provide a concise visual representation of these features and their characteristics across different scales, aiding in the exploration and analysis of complex data sets. A more formal explanation of these concepts is discussed in Section 2.1.

2.1. Persistent Homology

We discuss some of the main homology groups used for data analytics. We start by presenting the *Persistent Homology Group* or Persistent Homology, PH.

PH identifies topological features of a space at different scales. Features that are consistently detected across a broad range of scales are considered more likely to be true features of the space rather than being influenced by factors such as sampling errors or noise. To use persistent homology, the space must be represented as a *simplicial complex* (i.e., a set of *polytopes*, like points, line segments, triangles, tetrahedra and so on) and a *filtration*, or a nested sequence of increasing subsets, must be defined using a distance function on the space.

To clarify such abstract concepts as simplicial complex and filtration, let us consider a set of measurements of some property (or properties) of interest in terms of the associated features for each point (see Figure 2A). We customarily call this set a *point cloud;* this represents the data. Point clouds, which, as we said, are simply collections of points, do not have many *interesting* topological properties per se. However, we can analyze their topology by placing a ball of radius ϵ around each point. This method (called filtration) allows for us to encode geometric information by increasing the value of ϵ, which determines how much the boundaries of the shape blur and expand. As ϵ increases (Figure 2B–E), points that were initially distinct may begin to overlap, altering the concept of proximity. Essentially, this process involves taking, let us say, *an impressionist* painting (when we partially close our eyes to reveal details, like we do to appreciate a picture from Claude Monet) looking at the point cloud to offer it a more defined shape.

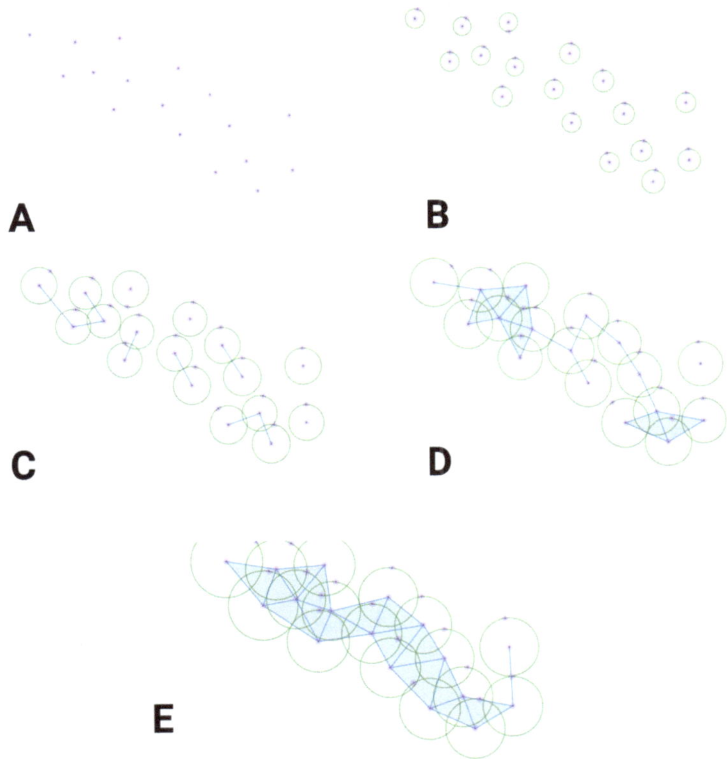

Figure 2. Persistent homology via FSC. Panel (**A**) presents a set of data points in feature space. In panel (**B**), we put a ball of radius ϵ around each data point. In panels (**B–E**), we increase radius ϵ. Neighborhoods start to overlap, giving rise to the establishment of a filtered simplicial complex.

A bit more formally, a simplicial complex is a collection of finite sets of points, called vertices, that are connected by edges, line segments connecting two vertices, and faces, which are polygons with three or more edges. The vertices, edges, and faces of a simplicial complex must satisfy certain conditions:

1. Every face of the complex must be a simplex, that is, a triangle or a higher-dimensional analogue of a triangle.
2. Every face of the complex must be a subset of one of the vertices of the complex.
3. If a face of the complex is a subset of another face, then the larger face must be a subset of one of the vertices of the complex.

Once we learn to build a simplicial complex for a given scale (value of ϵ) by changing the value of ϵ, what we do is create a *filtered simplicial complex* (FSC). Every topological property (such as the homologies H_k) that persists through the FSC is a PH. Intuitively, different phenomena under study give rise to different point clouds that, when analyzed via an FSC, have different PHs.

2.1.1. Building the FSC

By building the PH to a given point cloud, one aims to create a complex that approximates the original manifold using the given points. To achieve this, connections are established between points by adding edges between pairs of points; faces between triples of points; and so on. To determine which connections to create, we introduce a parameter called the filtration value (the ϵ we already mentioned), which limits the maximum length of the edges that can be included in our simplices. We vary ϵ and build the complex at each value, calculating the homology of the complex at each step.

There are three main strategies for using ϵ to assign simplices: the *Vietoris–Rips* strategy, the *witness* strategy, and the *lazy-witness* strategy [9]. The Vietoris–Rips strategy adds an edge between two points if their distance is less than ϵ, a face between three points if their pairwise distance is less than ϵ, and so on. This approach is accurate but computationally expensive. The witness strategy uses two sets of points, called *landmark points* and *witness points*, to create the complex. Landmark points are used as vertices and edges are added between two landmark points if there is a witness point within distance ϵ of both points, faces are added if there is a witness point within ϵ of all three points, and so on. The lazy-witness strategy is similar to the witness strategy in terms of how edges are assigned, but simplices of higher order are added anywhere there are n points that are all connected by edges.

2.1.2. Calculating the PH

Once we choose a filtration, it is possible to calculate the homology groups (the H_k's) of each space in the filtration. Homology groups are a way of describing the topological features of a space, such as connected components, holes, and voids. Depending on the particular task, we may choose a maximum value of k to build the first k homology groups. Then, we can use these homology groups to create a *barcode* or *persistence diagram*, which shows how the topological features of the space change as the scale changes [9].

To calculate persistent homology, it is possible to use a variety of algorithms, such as the already mentioned Vietoris–Rips complex, the Čech complex, or the alpha complex. These algorithms construct a simplicial complex from the data, which can then be used to calculate the homology groups of the space.

Calculating persistent homology and interpreting the results is a non-trivial task. Several issues need to be considered and decisions need to be taken in every step of the process. We can summarize the process as follows:

1. **Simplicial Complex Construction**: Begin by constructing a simplicial complex from your data. This complex can be based on various covering maps, such as the Vietoris–Rips complex, yhe Čech complex, or the alpha complex, depending on the chosen strategy (see Sections 2.4 and 2.5, as well as Table 1 below).
   The simplicial complex consists of vertices (0—simplices), edges (1—simplices), triangles (2—simplices), and higher-dimensional simplices. The choice of the complex depends on user data and the topological features of interest.
2. **Filtration**: Introduce a filtration parameter (often denoted as ϵ) that varies over a range of values. This parameter controls which simplices are included in the complex based on some criterion (e.g., distance threshold).
   As ϵ increases, more simplices are added to the complex, and the complex evolves. The filtration process captures the topological changes as ϵ varies.
3. **Boundary Matrix**: For each value of ϵ in the filtration, compute the boundary matrix (also called the boundary operator) of the simplicial complex. This matrix encodes the relations between simplices.
   Each row of the boundary matrix corresponds to a $(k-1)$-dimensional simplex, and each column corresponds to a k-dimensional simplex. The entries indicate how many times a $(k-1)$-dimensional simplex is a face of a k-dimensional simplex.
4. **Persistent Homology Calculation**: Perform a sequence of matrix reductions (e.g., Gaussian elimination) to identify the *cycles* and *boundaries* in the boundary matrix.
   A cycle is a collection of simplices whose boundaries sum to zero, while a boundary is the boundary of another simplex.
   Persistent homology focuses on tracking the birth and death of cycles across different values of ϵ. These births and deaths are recorded in a persistence diagram or a barcode.

5. **Persistence Diagram or Barcode**: The persistence diagram is a graphical representation of the births and deaths of topological features (connected components, loops, voids) as ϵ varies.
   Each point in the diagram represents a topological feature and is plotted at birth (x-coordinate) and death (y-coordinate).
   Interpretation:
   A point in the upper-left quadrant represents a long-lived feature that persists across a wide range of ϵ values.
   A point in the lower-right quadrant represents a short-lived feature that exists only for a narrow range of ϵ values.
   The diagonal represents features that are consistently present throughout the entire range of ϵ values.
   The distance between the birth and death of a point in the diagram quantifies the feature's *persistence* or *lifetime*. Longer persistence indicates a more stable and significant feature.
6. **Topological Summaries**: By examining the persistence diagram or a barcode, information can be extracted about the prominent topological features in a user data set.
   Features with longer persistence are considered more robust and significant.
   The number of connected components, loops, and voids can be quantified by counting points in specific regions of the diagram.

2.2. The Mapper Algorithm

Mapper is a recently developed algorithm that provides a reliable tool for topological data analysis. Mapper allows for researchers to identify and visualize the structure of a data set by creating a graph representation of the data [10].

The Mapper algorithm consists in the following steps [11] (see Figure 3):

1. Covering the data set: The original data set (Figure 3a) is partitioned into a number of overlapping subsets called *nodes* (Figure 3b). This is accomplished using a function called the *covering map*. The covering map assigns each point in the data set to a node. Since the nodes are allowed to overlap, every point potentially belongs to multiple nodes.
   There are several different ways to define a covering map. The choice of a covering map, however, can significantly affect the resulting Mapper graph. Some common approaches to define a covering map include:
   (a) Filtering: The data set is partitioned based on the values of one or more variables. A data set may, for instance, be partitioned based on the values of a categorical variable, such as gender or race.
   (b) Projection: Data set partitioning is performed by calculating the distance between points in the data set and using it as a membership criteria. This can be achieved using a distance function such as Euclidean distance or cosine similarity.
   (c) Overlapping intervals: The data set is partitioned into overlapping intervals, such as bins or quantiles. This can be useful for data sets that are evenly distributed or those having a known underlying distribution.

   The choice of covering map depends on the characteristics of the data set and the research question being addressed. It is important to choose a covering map that is appropriate for the data set and that will yield meaningful results.
2. Clustering the nodes: The nodes are then clustered using a clustering algorithm, such as k-means or single-linkage clustering. The resulting clusters (Figure 3c) represent the topological features of the data set, and the edges between the clusters represent the relationships between the features.

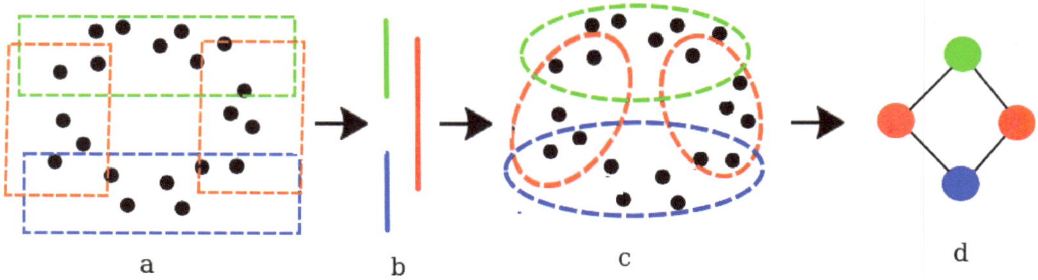

Figure 3. The steps of the Mapper algorithm. (**a**) The data, a cloud of points. (**b**) The projection of the data into a lower dimension space. (**c**) The preimage is clustered and (**d**) a graph is built based on the clustered groups. See text.

The resulting graph (Figure 3d), called a *Mapper graph*, can be used to identify patterns and relationships in the data set that may not be apparent from other forms of visualization [12].

2.3. Multidimensional Scaling

In the context of topological data analysis, multidimensional scaling (MDS) is a method to visualize the relationships between a set of complex objects as projected in a lower-dimensional space. MDS is a versatile technique used in various fields for analyzing and visualizing relationships between objects. It can be used both in classic data analysis approaches and in conjunction with TDA methods. MDS works by creating a map of the objects in which the distance between said objects reflects (to a certain point) the dissimilarity between them [13]. MDS is often used along other techniques, like clustering, to analyze patterns in data sets that have a large number of variables. Multidimensional scaling can help identify relationships between objects that are not immediately apparent; hence, it is useful to visually explore complex data sets [14]. There are several different algorithms for performing MDS, including classical MDS, nonmetric MDS, and metric MDS.

Classical MDS is the most common method (see Figure 4). In a nutshell, we start with a set of n points in a space of high dimension (m), then we introduce a measure of similarity (or dissimilarity), for instance, a distance (such as the Euclidean distance), then we have a square symmetric matrix with $n \times n$ pairwise distances, and MDS is attained by performing Principal Coordinate Analysis (i.e., eigenvalue decomposition) of such matrix. The result is a set of lower-dimensional coordinates for n points. Hence, classic MDS is based on the idea of preserving the pairwise distances between objects in the projected low-dimensional map. Classical MDS finds the map that best preserves the distances between objects using an optimization algorithm. The Nonmetric MDS method is similar to classical MDS, but it does not assume that the dissimilarities between objects are metric—i.e., represented by a continuous scale; instead, it preserves the rank order of the dissimilarities between objects, instead of the absolute values. Metric MDS, conversely, is a variant of classical MDS based on *stress minimization*, that is, by considering the difference between the distances in the low-dimensional map and the dissimilarities in the data set. This method is used when the dissimilarities between objects can be represented by a continuous scale.

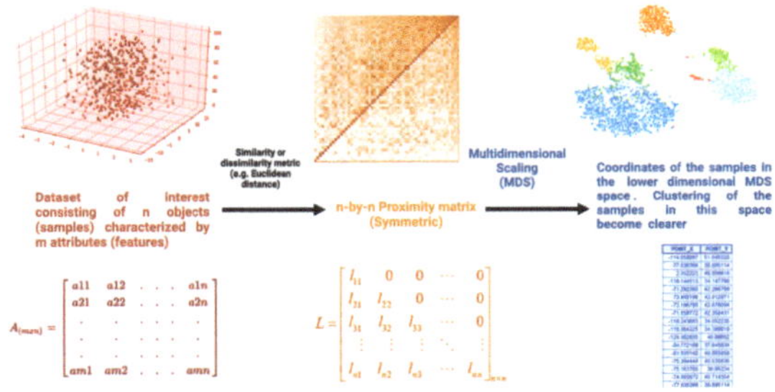

Figure 4. Classic multidimensional scaling.

In general, classical MDS is the most widely used method, but nonmetric MDS and metric MDS may be more appropriate in certain situations.

How to Determine the Scaling Approach?

The choice between classical MDS, nonmetric MDS, and metric MDS depends ultimately on the characteristics of the data and the specific research question. Some general guidelines are as follows:

1. Classical MDS:
    - **When to Use:** Classical MDS is suitable when using metric (distance) data that accurately represent the pairwise dissimilarities between objects. In classical MDS, the goal is to find a configuration of points in a lower-dimensional space (usually 2D or 3D) that best approximates the given distance matrix.
    - **Pros:** It preserves the actual distances between data points in lower-dimensional representation.
      It provides a faithful representation when the input distances are accurate.
      It is well-suited for situations where the metric properties of the data are crucial.
    - **Cons:** It assumes that the input distances are accurate and may not work well with noisy or unreliable distance data.
      It may not capture the underlying structure of the data if the metric assumption is violated.
2. Nonmetric MDS:
    - **When to Use:** Nonmetric MDS is appropriate when using ordinal or rank-order data, where the exact distances between data points are not known, but their relative dissimilarities or rankings are available. Nonmetric MDS finds a configuration that best preserves the order of dissimilarities.
    - **Pros:** It is more flexible than classical MDS and can be used with ordinal data.
      It can handle situations where the exact distances are uncertain or difficult to obtain.
    - **Cons:** It does not preserve the actual distances between data points, so the resulting configuration is only an ordinal representation.
      The choice of a monotonic transformation function to convert ordinal data into dissimilarity values can affect the results.
3. Metric MDS:
    - When to Use: Metric MDS can be used when using data that are inherently non-metric, but it is believed that transforming it into a metric space could reveal meaningful patterns. Metric MDS aims to find a metric configuration that best approximates the non-metric dissimilarities.
    - **Pros:** It provides a way to convert non-metric data into a metric representation for visualization or analysis.
      It can help identify relationships in the data that may not be apparent in the original non-metric space.

- **Cons:** The success of metric MDS depends on the choice of the transformation function to convert non-metric data into metric distances.
  It may not work well if the non-metric relationships in the data are too complex or cannot be adequately approximated by a metric space.

In summary, the choice between classical, nonmetric, and metric MDS depends on the nature of user data and the goals of analysis. If metric data are accurate and preservation of the actual distances is desired, classical MDS is appropriate. If data are ordinal or dissimilarity measures are uncertain, nonmetric MDS may be more suitable. Metric MDS can be considered when it is desired to convert non-metric data into a metric space for visualization or analysis, but it requires careful consideration of the transformation function.

2.4. Choosing the Covering Map

When choosing a covering map in TDA, there are several characteristics of the data sets which are relevant to consider; among these, we can mention the following:

1. **Data Dimensionality:** The dimensionality of the data under consideration is crucial. Covering maps should be chosen to preserve the relevant topological information in the data. For high-dimensional data, dimension reduction techniques may be applied before selecting a covering map.
2. **Noise and Outliers:** The presence of noise and outliers in the data can affect the choice of a covering map. Robust covering maps can help mitigate the influence of noise and outliers on the topological analysis.
3. **Data Density:** The distribution of data points in the feature space matters. A covering map should be chosen to account for variations in data density, especially if there are regions of high density and regions with sparse data.
4. **Topological Features of Interest:** It is important to consider the specific topological features one is interested in analyzing. Different covering maps may emphasize different aspects of data topology, such as connected components, loops, or voids. The election of a covering map should align with particular research objectives.
5. **Computational Efficiency:** The computational complexity of calculating the covering map should also be taken into account. Some covering maps may be computationally expensive, which can be a limiting factor for large data sets.
6. **Continuous vs. Discrete Data:** It should be determined whether the data under analysis are continuous or discrete. The choice of a covering map may differ based on the nature of the data.
7. **Metric or Non-Metric Data:** Some covering maps are designed for metric spaces, where distances between data points are well defined, while others may work better for non-metric or qualitative data.
8. **Geometric and Topological Considerations:** The geometric and topological characteristics of user data should be considered. Certain covering maps may be more suitable for capturing specific geometric or topological properties, such as persistence diagrams or Betti numbers.
9. **Domain Knowledge:** Domain-specific knowledge should be incorporated into user choice of a covering map. Understanding the underlying structure of the data can guide the user in selecting an appropriate covering map.
10. **Robustness and Stability:** The robustness and stability of the chosen covering map shpuld be assessed. TDA techniques should ideally produce consistent results under small perturbations of the data or variations in sampling.

In practice, there are various covering maps and TDA algorithms available, such as Vietoris–Rips complexes, Čech complexes, and alpha complexes. The choice of covering map should be guided by a combination of these factors, tailored to the specific characteristics and goals of user data analysis. It may also involve some experimentation to determine which covering map best captures the desired topological features.

Table 1. Some types of covering maps and their main applications.

Covering Map	Type of Data	Brief Description
Vietoris–Rips Complex	Point cloud data, particularly when dealing with metric spaces. It is often used in applications like sensor networks, molecular chemistry, and computer graphics	The Vietoris–Rips complex connects points in the data if they are within a certain distance (the radius parameter) of each other, forming simplices (e.g., edges, triangles, tetrahedra) based on pairwise distances
Čech Complex	Similar to the Vietoris–Rips complex, Čech complexes are used for point cloud data in metric spaces	The Čech complex connects points if they belong to the same open ball of a specified radius. It can capture similar topological features as the Vietoris–Rips complex but may have a different geometric structure
Alpha Complex	Alpha complexes are useful for point cloud data in metric spaces and provide an alternative representation of the topological structure	The alpha complex connects points with a Delaunay triangulation, considering balls whose radii can vary at each point to ensure that the complex is a subcomplex of the Vietoris–Rips complex
Witness Complex	Witness complexes are used for point cloud data but are particularly useful when dealing with data that may not be uniformly sampled or when dealing with non-metric or qualitative data	Witness complexes are constructed by selecting a subset of witness points from the data. Each witness point witnesses the presence of other data points within a specified distance. This can be used to capture topological features in a more robust way, especially when data is irregular
Mapper	Mapper is a flexible approach that can be applied to various types of data, including both metric and non-metric spaces	Mapper is not a traditional covering map but rather a method for creating a topological summary of data by combining clustering and graph theory. It can be adapted to different data types and is useful for exploratory data analysis
Rips Filtration and Čech Filtration	These are extensions of the Vietoris–Rips and Čech complexes, respectively, that allow for the analysis of topological features at different scales	By varying the radius parameter continuously, Rips filtration and Čech filtration produce a sequence of simplicial complexes. This can be useful for capturing topological features at different levels of detail and studying persistence diagrams

Different types of covering maps are thus best suited for different kinds of data. Some of the main covering maps used in TDA are presented in Table 1.

Ultimately, the choice of covering map depends on the specific characteristics of user data, such as dimensionality, metric properties, and the topological features of interest. It is often beneficial to experiment with different covering maps and parameters to determine which one best captures the desired topological information for a particular data set. Additionally, combining multiple covering maps and TDA techniques can provide a more comprehensive understanding of complex data sets.

2.5. Different Strategies for Topological Feature Selection

The Vietoris–Rips strategy, the witness strategy, and the lazy-witness strategy are some of the best-known TDA methods to capture and analyze the topological features of data sets. Each of these strategies has its own advantages and disadvantages.

2.5.1. Vietoris–Rips (VR) Strategy

The main advantage of the VR strategy is its simplicity, since the VR complex is relatively easy to understand and implement. It connects data points based on a fixed distance threshold, which is quite intuitive. Another advantage of the VR lies on its widespread use, for there is a significant body of literature and software implementations available. It works well with data in metric spaces where distances between points are well defined.

One of the disadvantages is that VR is quite sensitive to certain parameters; the choice of distance threshold (radius parameter) can significantly impact the topology of the resulting complex. Selecting an appropriate threshold can be challenging and may require prior knowledge of the data. VR can be also challenging for its computational burden: constructing the VR complex can be computationally expensive, especially for large data sets or high-dimensional data. VR is also limited in terms of robustness: the VR complex is sensitive to noise and outliers, and small perturbations in the data can lead to significant changes in complex topology.

2.5.2. Witness Strategy (WS)

The witness strategy (WS), in turn, is more robust to noise and outliers compared to the VR complex. It selects a subset of *witness points* that can capture the topology of the data more effectively. WS is more flexible; witness complexes can be applied to both metric and non-metric data, making them versatile for various data types and able to handle data with varying sampling densities; this, they are suitable for irregularly sampled data sets.

Implementation of WS, however, can be more involved than that of the VR complex, as it requires selecting witness points and computing their witness neighborhoods. Also, while witness complexes are more robust, they still depend on parameters like the number of witness points and the witness radius. Choosing appropriate parameters can indeed be a non-trivial task.

2.5.3. Lazy-Witness Strategy (LW)

The LW strategy is an optimization of the witness strategy that reduces computational cost. It constructs witness complex *on-the-fly* as needed, potentially saving memory and computation time. Like WS, the LW strategy is robust to noise and outliers.

In spite of these advantages, there are also shortcomings. Implementing the LW strategy can be more complex than implementing the basic witness strategy, as it requires careful management of data structures and computational resources. While it can be more memory efficient than precomputing a full witness complex, the LW strategy still consumes memory as it constructs the complex in real time. This may still be a limitation for very large data sets.

In summary, the choice between the Vietoris–Rips strategy, witness strategy, and lazy-witness strategy depends on the specific characteristics of user data and the computational resources available. The Vietoris–Rips complex is straightforward but sensitive to parameter choice and noise. The witness strategy offers improved robustness but may require more effort in parameter tuning. The lazy-witness strategy combines robustness with some memory and computation efficiency, making it a good choice for large data sets. Experimentation and a deep understanding of user data characteristics are essential when selecting the most appropriate strategy for user TDA analysis.

3. Applications of TDA to Analyze Cardiovascular Signals

3.1. General Features

Aljanobi and Lee [5] recently applied the Mapper algorithm to predict heart disease. They selected nine significant features in each of the two UCI heart disease data sets (Cleveland and Statlog). The authors then used a tri-dimensional SVD filter to improve the filtering process. As a result, they observed an accuracy of 99.32% in the Cleveland data set and 99.62% in the Statlog data set in predicting heart disease.

Though not precisely *signals* but rather contextual string corpora (i.e., *texts*), TDA has also been applied to the analysis of structured and unstructured text in EHRs and clinical notes. This is achieved by first converting textual data into a suitable format for analysis. This can involve techniques such as tokenization, stemming, and removing stop words. Afterwards, one needs to represent the text as numerical vectors using methods like Term Frequency-Inverse Document Frequency (TF-IDF) or word embeddings (Word2Vec, GloVe). Once data are vectorized and tokenized, TDA can be applied. Lopez and coworkers [15], for instance, used the Mapper algorithm to classify distinctive subsets of patients receiving optimal treatments post-acute myocardial infarction (AMI) in order to identify high-risk subgroups of patients for having a future adverse event (AE) such as death, heart failure hospitalization, or recurrent myocardial infarction. A retrospective analysis of 31 clinical variables from the EHR of 798 AMI subjects was conducted at a single center. The subjects were divided into high- and low-risk groups based on their probability of survival without AEs at 1 year. TDA identified six subgroups of patients. Four of these subgroups, totaling 597 subjects, had a probability of survival without AEs that was greater than a one-fold change, and were considered low risk. The other two subgroups, totaling 344 subjects, had a probability of survival without AEs that was less than a one-fold change, and were considered high risk. However, 143 subjects (18% of the total) were classified as intermediate risk because they belonged to both high- and low-risk subgroups. TDA was also able to significantly stratify AMI patients into three subgroups with distinctive rates of AEs up to 3 years after AMI. This approach to EHR-based risk stratification does not require additional patient interaction and is not dependent on prior knowledge, but more studies are needed before it can be used in clinical practice.

3.2. ECG Data and Heart Rate Signals

In another recent study, Yan and coworkers [16] explored the use of topological data analysis to classify electrocardiographic signals and detect arrhythmias. Cardiac arrhythmias are abnormal heart rhythms or irregularities in the heartbeat that may be too fast (tachycardia), too slow (bradycardia), or irregular. Arrhythmias may originate from problems with the heart's electrical system, damage to the heart muscle, or other medical conditions. The most common types of arrhythmias are *Atrial Fibrillation* (AF) defined as an irregular and often rapid heartbeat that can lead to stroke and other heart-related complications; *Atrial Flutter* which is similar to AF, is characterized by a rapid, regular heartbeat originating in the atria; *Supraventricular Tachycardia* is characterized by episodes of rapid heart rate originating above the heart's ventricles; *Ventricular Tachycardia*, a fast, regular beating of the heart's lower chambers (ventricles) that can be life-threatening; *Ventricular Fibrillation* (VF) which is a chaotic, rapid heartbeat that can be life-threatening and requires immediate medical attention; and finally *Bradycardia*, a slower than normal heart rate, often caused by issues with the heart's natural pacemaker.

In the particular case of reference [16], phase space reconstruction was used to convert the signals into point clouds, which were then analyzed using topological techniques to extract persistence landscapes as features for the classification task. The authors found that the proposed method was effective, with a normal heartbeat class recognition rate of 100% when using just 20% of the training set, and recognition rates of 97.13% for ventricular beats, 94.27% for supraventricular beats, and 94.27% for fusion beats. This ability to maintain high performance with a small training sample space makes TDA particularly suitable for personalized analysis.

One particularly difficult problem in cardiovascular disease diagnostics with important implications for therapy is the evolution of atrial fibrillation [17]. Indeed, the progression of AF from paroxysmal to persistent or permanent forms has become a significant issue in cardiovascular disorders. Information about the pattern of presentation of AF (paroxysmal, persistent, or permanent) is useful in the management of algorithms for each category, which aims to reduce symptoms and prevent severe problems associated with AF. Until now, AF classification has been based on the duration and number of episodes. In particular, changes in complexity of Heart Rate Variation (HRV) may contain clinically relevant signals of impending systemic dysregulation. HRV measures the fluctuations in the time intervals between consecutive heartbeats, providing insights into the autonomic nervous system's activity, particularly the balance between its sympathetic and parasympathetic branches. A number of nonlinear methods based on phase space and topological properties can provide further insight into HRV abnormalities such as fibrillation. In an effort to provide a tool for the qualitative classification of AF stages, Safarbaly and Golpayegani [18] proposed two geometrical indices (fractal dimension and persistent homology) based on HRV phase space, which were able to successfully replicate the changes in AF progression.

Their studied population included 38 lone AF patients and 20 normal subjects, whose data were collected from the Physio-Bank database [19]. "Time of Life (TOL)" was proposed as a new feature based on the initial and final Čech radius in the persistent homology diagram. A neural network was implemented to demonstrate the effectiveness of both TOL and fractal dimension as classification features, resulting in a classification accuracy of 93%. The proposed indices thus provide a signal representation framework useful for understanding the dynamic changes in AF cardiac patterns but also for classifying normal and pathological rhythms.

PH was also used by Graff et al. to study HRV [20]; the authors suggested the use of persistent homology for the analysis of HRV, relying on some topological descriptors previously used in the literature and introducing new ones that are specific to HRV, later discussing their relationship to standard HRV measures. The authors showed that this novel approach produces a set of indices that may be as useful as classical parameters in distinguishing between series of beat-to-beat intervals (RR-intervals) in healthy individuals as well as in patients who have experienced a stroke.

Also in the context of fibrillation (this time for the prediction of early ventricular fibrillation), Ling and coworkers [21] proposed a novel feature based on topological data analysis to increase the accuracy of early VF prediction. In their work, the heart activity was first treated as a cardiac dynamical system, which was described through phase space reconstruction. The topological structure of the phase space was then characterized using persistent homology, and statistical features of the topological structure were extracted and defined as TDA features. To validate the prediction performance of the proposed method, 60 subjects (30 VF, 30 healthy) from three public ECG databases were used. The TDA features showed a superior accuracy of 91.7% compared to heart rate variability features and box-counting features. When all three types of features were combined as fusion features, the optimal accuracy of 95.0% was achieved. The fusion features were then ranked, and the first seven components were all from TDA features. The proposed features may have a significant effect on improving the predictive performance of early VF.

A similar approach was taken by Mjahad, et al. [22], who applied TDA to generate novel features contributing to improve both detection and classification performance of cardiac arrhythmias such as Ventricular Fibrillation (VF) and Ventricular Tachycardia (VT). The electrocardiographic (ECG) signals used for this evaluation were obtained from the standard MIT-BIH and AHA databases. The authors evaluated two types of input data for classification: TDA features and the so-called Persistence Diagram Image (PDI). When using the reduced TDA-derived features, a high average accuracy of nearly 99% was observed to discriminate between four types of rhythms (98.68% for VF; 99.05% for VT; 98.76% for normal sinus; and 99.09% for other rhythms), with specificity values higher than 97.16% in all cases. In addition, a higher accuracy of 99.51% was obtained when discriminating between shockable (VT/VF) and non-shockable rhythms (99.03% sensitivity and 99.67% specificity). These results show that the use of TDA-derived geometric features, combined with the k-Nearest Neighbor (kNN) classifier, significantly improves classification performance compared to previous works. These results were achieved without pre-selection of ECG episodes, suggesting that these features may be successfully used in Automated External Defibrillation (AED) [23,24] and Implantable Cardioverter Defibrillation (ICD) [25,26] therapies.

Jiang and collaborators studied non-invasive atrial fibrillation using TDA on ballistocardiographic (BCG) data [27]. BCG refers to the measurement and recording of the ballistic forces generated by the ejection of blood from the heart during each cardiac cycle. These forces produce subtle vibrations and movements in the body, particularly in the chest and torso. Ballistocardiography is a non-invasive technique used to capture and analyze these mechanical movements, providing valuable information about cardiac function and hemodynamics. In this research, BCG series was transformed into a high-dimensional point cloud in order to capture more rhythmic information. These point clouds were then analyzed using TDA, resulting in persistent homology barcodes. The statistics of these barcodes were extracted as nine persistent homology features to quantitatively describe the barcodes. In order to test the effectiveness of this method for detecting atrial fibrillation (AF), the researchers collected BCG data from 73 subjects with both AF and non-AF segments, and applied six machine learning classifiers. The combination of these 9 features with 17 previously proposed features resulted in a 6.17% increase in accuracy compared to using 17 features alone ($p < 0.001$), with an overall accuracy of 94.50%. By selecting the most effective features using feature selection, the researchers were able to achieve a classification accuracy of 93.50%. According to the authors, these results suggest that the proposed features can improve AF detection performance when applied to a large amount of BCG data with diverse pathological information and individual differences.

TDA can also be combined with other data analytics/Ml approaches such as random forests (RF). Ignacio and collaborators developed a topological method to *inform* RF-based ECG classifiers [28]. In brief, in this approach, a two-level random forest model is trained to classify 12-lead ECGs using mathematically computable topological signatures as proxy for features informed by medical expertise. ECGs are treated as multivariate time series data and transformed into point cloud embeddings that capture both local and global structures and encode periodic information as attractor cycles in a high-dimensional space. Topological data analysis is then used to extract topological features from these embeddings, and these features are combined with demographic data and statistical moments of RR intervals calculated using the Pan–Tompkins algorithm for each lead to train the classifier. This multi-class classification task aims to leverage the medical expertise represented in the topological signatures to accurately classify ECGs.

The same group combined TDA and ML approaches to study atrial fibrillation in ECGs [29], showing that topological features can be used to accurately classify single-lead ECGs by applying delay embeddings to map ECGs onto high-dimensional point clouds, which convert periodic signals into algebraically computable topological signatures, thus allowing them the use of these topological features to classify ECGs.

A similar approach coupling TDA with Deep Learning to study ECGs has been recently developed [30]. Training deep learning models on time series data such as ECG poses some challenges such as lack of labeled data and class imbalance [31,32]. The authors used TDA to improve the performance of deep learning models on this type of data. The authors found that using TDA as a time-series embedding method for input to deep learning models resulted more effective than training these models directly on raw data. TDA in this context serves as a generic, low-level feature extractor that can capture common signal patterns and improve performance with limited training data. Experiments on public human physiological biosignal data sets showed that this approach leads to improved accuracy, particularly for imbalanced classes with only a few training instances compared to the full data set.

Conventional TDA of ECG time series can be further improved by considering the time delay structure to generate higher dimensional mappings of the original series [33] able to be analyzed via TDA. This was the approach taken by Fraser and coworkers [34], who found that TDA visualizations are able top unveil ectopic and other abnormal occurrences in long signals, indicating a promising direction for the study of longitudinal physiological signals.

A different approach to deal with ECG time series using TDA makes use of optimal representative cycles. In reference [35], the authors applied a topological data-analytic method to identify parts of an electrocardiogram (ECG) signal that are representative of specific topological features and proposed that these parts correspond to the P, Q, S, and T-waves in the ECG signal. They then used information about these parts of the signal, identified as P, Q, S, and T-waves, to measure PR-interval, QT-interval, ST-segment, QRS-duration, P-wave duration, and T-wave duration. The method was tested on simulated and real Lead II ECG data, demonstrating its potential use in analyzing the morphology of the signal over time and in arrhythmia detection algorithms.

3.3. Stenosis and Vascular Data

TDA has also been used to study stenosis, an abnormal narrowing or constriction of blood vessels. In the cardiovascular system, arterial stenosis can be particularly significant. Atherosclerosis, a condition characterized by the buildup of plaque on the inner walls of arteries, is a common cause of arterial stenosis. Nicponski and collaborators [36] demonstrated the use of persistent homology to assess the severity of stenosis in different types of stenotic vessels. They introduced the concept of *critical failure value*, which applies one-dimensional homology to these vessels as a way to quantify the degree of stenosis. They also presented the spherical projection method, which could potentially be used to classify various types and levels of stenosis, and showed that the two-dimensional homology of the spherical projection can serve as a new index for characterizing blood vessels. It is worth noticing that, as in many other instances in data analytics, data pre-processing often represents a crucial stage [37].

3.4. TDA in Echocardiography

Applications of TDA to echocardiographic data have also been developed. Such is the case of the work the group of Tokodi [38,39] who analyzed a cohort of 1334 patients to identify similarities among patients based on several echocardiographic measures of left ventricular function. Echocardiography, a medical imaging technique that uses sound waves to create detailed images of the heart, generates 2D and 3D maps of vascular structure. However, accurate image reconstruction is far from trivial, in particular for small convoluted cavities. A network was developed in reference [38] to represent these similarities, and a group classifier was used to predict the location of 96 patients with two consecutive echocardiograms in this network. The analysis revealed four distinct regions in the network, each with significant differences in the rate of major adverse cardiovascular event (MACE) rehospitalization. Patients in the fourth region had more than two times the risk of MACE rehospitalization compared to those in the other regions. Improvement or stability in Regions I and II was associated with lower MACE rehospitalization rates compared to worsening or stability Regions III and IV. The authors concluded that TDA-driven patient similarity analysis may improve the precision of phenotyping and aid in the prognosis of patients by tracking changes in cardiac function over time.

4. Conclusions

We showed that within the realm of cardiovascular signal analysis, the utilization of Topological Data Analysis (TDA) displayed remarkable promise across diverse applications (see Table 2). As we mentioned, Aljanobi and Lee applied the Mapper algorithm to predict heart disease with exceptional accuracy, achieving 99.32% accuracy in the Cleveland data set and 99.62% in the Statlog data set. This underscores the effectiveness of TDA in identifying significant features for disease prediction. Moving beyond disease prediction, TDA proved invaluable in the risk stratification of patients post-acute myocardial infarction (AMI). Lopez and colleagues employed the Mapper algorithm on Electronic Health Records (EHR), successfully classifying distinct subsets of AMI patients. The insights gained into high-risk subgroups and their susceptibility to adverse events up to 3 years post AMI showcase the potential of TDA in personalized patient care.

The application of TDA to electrocardiographic (ECG) signal classification yielded notable outcomes. Yan and Ling explored TDA for arrhythmia detection and early ventricular fibrillation prediction, respectively. Yan's use of persistent homology achieved high recognition rates, while Ling's approach demonstrated superior accuracy compared to traditional features, indicating the robustness of TDA in diverse ECG applications. Safarbaly and Golpayegani delved into the progression of atrial fibrillation (AF) using persistent homology. Their novel geometric indices, based on HRV phase space, successfully replicated changes in AF stages, attaining a classification accuracy of 93%. This not only sheds light on AF dynamics, but also presents a potential tool for qualitative AF classification.

The synergy of TDA with machine learning was evident in Mjahad's work, where TDA-derived geometric features significantly improved the classification of cardiac arrhythmias. The application extended to non-invasive atrial fibrillation detection by Jiang and collaborators, showcasing an enhanced accuracy compared to traditional features alone. Byers and team explored the integration of TDA with deep learning for ECG classification. Using TDA as a time-series embedding method for deep learning models resulted in improved accuracy, particularly for imbalanced classes, addressing challenges in classification tasks with limited training data. TDA's versatility extended to stenosis severity assessment in various vessels, as demonstrated by Nicponski and collaborators. The critical failure value and spherical projection methods provided quantifiable measures, highlighting the potential of TDA in characterizing blood vessels. Lastly, TDA was applied to echocardiographic data by Tokodi and team for patient similarity analysis. The Mapper algorithm revealed distinct regions in a network representing similarities among patients based on echocardiographic measures. The identified regions were associated with significant differences in major adverse cardiovascular events (MACE) rehospitalization rates, offering insights into patient prognosis.

In conclusion, the diverse applications of TDA in cardiovascular signal analysis underscore its effectiveness in disease prediction, risk stratification, signal classification, and patient similarity analysis. The integration of TDA with machine learning and deep learning techniques presents exciting avenues for advancing our understanding of complex cardiovascular patterns and improving clinical outcomes (for software implementations see Table A1). However, further research and validation are essential to solidify the clinical applicability of these TDA-based approaches.

Table 2. Comparison of TDA Models in Cardiovascular Signal Applications.

Study	Model Used	Application	Data Set	Features/Methods	Accuracy/Performance
Aljanobi and Lee [5]	Mapper Algorithm	Heart Disease Prediction	UCI Heart Disease Data sets (Cleveland, Statlog)	Nine significant features, tri-dimensional SVD filter	Cleveland: 99.32%, Statlog: 99.62%
Lopez et al. [15]	Mapper Algorithm	Risk Stratification of AMI Patients	EHR of 798 AMI subjects	Clinical variables from EHR, Mapper algorithm	3-year AE rates stratification
Yan et al. [16]	Persistent Homology	Arrhythmia Detection	ECG Signals	Phase space reconstruction, persistence landscapes	Normal: 100%, Ventricular: 97.13%, Supraventricular: 94.27%
Safarbaly and Golpayegani [18]	Persistent Homology	AF Progression	Physio-Bank Database	Fractal dimension, persistent homology	Classification accuracy: 93%
Graff et al. [20]	Persistent Homology	HRV Analysis	Healthy individuals and post-stroke patients	Topological descriptors, comparison with standard HRV measures	Distinction between RR-intervals
Ling et al. [21]	Persistent Homology	Early Ventricular Fibrillation Prediction	Public ECG Databases	Phase space reconstruction, persistent homology, statistical features	Accuracy: 95.0%
Mjahad et al. [22]	Persistent Homology	Arrhythmia Classification	MIT-BIH and AHA Databases	TDA features and Persistence Diagram Image (PDI)	Accuracy: 99%, Sensitivity: 99.03%, Specificity: 99.67%
Jiang et al. [27]	Persistent Homology	Non-invasive AF Detection	BCG Data from 73 subjects with AF and non-AF segments	TDA on BCG data, machine learning classifiers	Classification accuracy: 94.50%
Ignacio et al. [28]	TDA (Informed Random Forests)	ECG Classification	12-lead ECGs	Topological signatures, random forests	Not specified
Ignacio et al. [29]	TDA (Informed Random Forests)	Atrial Fibrillation Classification	Single-lead ECGs	Delay embeddings, topological features	Not specified
Byers et al. [30]	TDA combined with Deep Learning	ECG Classification	Public human physiological biosignal data sets	TDA as time-series embedding for deep learning models	Improved accuracy for imbalanced classes
Fraser et al. [34]	TDA with Time Delay Structure	ECG Visualization	Longitudinal physiological signals	Time delay structure, TDA visualizations	Unveiling abnormal occurrences
Dlugas et al. [35]	TDA with Optimal Representative Cycles	ECG Signal Morphology Analysis	Simulated and real Lead II ECG data	Identification of P, Q, S, T-waves, measurement of intervals	Not specified
Nicponski et al. [36]	Persistent Homology	Stenosis Severity Assessment	Various types of stenotic vessels	Critical failure value, spherical projection, 2D homology	Quantification of stenosis severity
Tokodi et al. [38]	Mapper Algorithm	Patient Similarity Analysis	Echocardiographic measures of left ventricular function	Network representation, group classifier	Prognosis of patients

Perspectives and Limitations

We presented a panoramic (and by necessity incomplete) view of the applications of topological data analysis to the study of cardiovascular signals. We must, however, stress that, as every analytic technique, TDA has a set of limitations and a range of applicability. Among its limitations, we can include its inherent complexity both to implement the analyses and to interpret the results. Also relevant is TDA's sensitivity to noise and reliance on assumptions of the data distribution (for instance, one assumes that filtration can be carried out unambiguously). Another aspect to consider is that topological data analysis can be computationally intensive, particularly for large or complex data sets. This, along potential biases in noisy signals, may preclude the accurate identification of the topological features of certain data sets, especially if the data are of low quality. As we said, topological data analysis relies on assumptions about the data set, for instance, the existence of a *natural* distance measure. This condition may not hold in all cases. This may potentially limit its applicability. Additionally, the results of topological data analysis can be difficult to interpret, particularly for non-experts, a fact that can make challenging to communicate the results of a topological analysis to a general audience. However, even in light of these limitations, TDA is a quite powerful approach that (perhaps in combination with other data analytic approaches) may result very useful for the study of complex biosignals, such as those arising in cardiology.

Author Contributions: Conceptualization: E.H.-L.; Investigation: E.H.-L., P.M., M.M.-G.; writing—original draft preparation: E.H.-L.; writing—review and editing: E.H.-L., P.M., M.M.-G. All authors have read and agreed to the published version of the manuscript.

Funding: This research received no external funding.

Institutional Review Board Statement: Not applicable.

Data Availability Statement: Not applicable.

Conflicts of Interest: The authors declare no conflicts of interest.

Appendix A. Computational Tools

In Table A1, we present some commonly used algorithms used to perform topological data analysis calculations implemented in libraries for common programming languages. Implementations vary in their capabilities, computational complexity and depth of documentation. These are, however, well documented enough, so that most people with scientific programming expertise may find them fairly usable and reliable. Where applicable, references to the underlying algorithms used are included.

Table A1. Computational tools for topological data analysis. All links were accesed on 4 June 2023.

Tool	Language	URL	Reference
TDA	R	https://cran.r-project.org/web/packages/TDA/index.html	[40,41]
TDAstats	R	https://cran.r-project.org/web/packages/TDAstats/index.html	[42]
TDApplied	R	https://cran.r-project.org/web/packages/TDApplied/index.html	
tdaverse	R	https://github.com/tdaverse/tdaverse	
PHAT	Python	https://bitbucket.org/phat-code/phat/src/master/	[43]
Dionysus 2	C++/Python	https://mrzv.org/software/dionysus2/	
GUDHI	C++/Python	https://gudhi.inria.fr/	
TTK	C++/Python	https://topology-tool-kit.github.io/	
DIPHA	C++/Matlab	https://github.com/DIPHA/dipha	[44]

Table A1. *Cont.*

Tool	Language	URL	Reference
Giotto-tda	Python	https://giotto-ai.github.io/gtda-docs/latest/library.html	
Jplex	Java	https://www.math.colostate.edu/~adams/jplex/index.html	
Ripser	C++	https://github.com/Ripser/ripser	
Ripser++	C++ CUDA	https://github.com/simonzhang00/ripser-plusplus	[45]
HERA	C++/Python	https://bitbucket.org/grey_narn/hera/src/master/	[46]
EIRENE	Julia	https://github.com/Eetion/Eirene.jl	
sklearn-tda	Python	https://github.com/MathieuCarriere/sklearn-tda	
Scikit-TDA	Python	https://github.com/scikit-tda	

References

1. Seetharam, K.; Shrestha, S.; Sengupta, P.P. Artificial intelligence in cardiovascular medicine. *Curr. Treat. Options Cardiovasc. Med.* **2019**, *21*, 242. [CrossRef] [PubMed]
2. Silverio, A.; Cavallo, P.; De Rosa, R.; Galasso, G. Big health data and cardiovascular diseases: A challenge for research, an opportunity for clinical care. *Front. Med.* **2019**, *6*, 36. [CrossRef]
3. Kagiyama, N.; Shrestha, S.; Farjo, P.D.; Sengupta, P.P. Artificial intelligence: Practical primer for clinical research in cardiovascular disease. *J. Am. Heart Assoc.* **2019**, *8*, e012788. [CrossRef] [PubMed]
4. Shameer, K.; Johnson, K.W.; Glicksberg, B.S.; Dudley, J.T.; Sengupta, P.P. Machine learning in cardiovascular medicine: Are we there yet? *Heart* **2018**, *104*, 1156–1164. [CrossRef] [PubMed]
5. Aljanobi, F.A.; Lee, J. Topological Data Analysis for Classification of Heart Disease Data. In Proceedings of the 2021 IEEE International Conference on Big Data and Smart Computing (BigComp), Jeju Island, Republic of Korea, 17–20 January 2021; pp. 210–213. [CrossRef]
6. Phinyomark, A.; Ibáñez-Marcelo, E.; Petri, G. Topological Data analysis of Biomedical Big Data. In *Signal Processing and Machine Learning for Biomedical Big Data*; CRC Press: Boca Raton, FL, USA, 2018; pp. 209–233.
7. Carlsson, G. The shape of biomedical data. *Curr. Opin. Syst. Biol.* **2017**, *1*, 109–113. [CrossRef]
8. Skaf, Y.; Laubenbacher, R. Topological data analysis in biomedicine: A review. *J. Biomed. Inform.* **2022**, *130*, 104082. [CrossRef]
9. Carlsson, G.; Vejdemo-Johansson, M. *Topological Data Analysis with Applications*; Cambridge University Press: Cambridge, UK, 2021.
10. Ristovska, D.; Sekuloski, P. Mapper algorithm and its applications. *Math. Model.* **2019**, *3*, 79–82.
11. Zhou, Y.; Chalapathi, N.; Rathore, A.; Zhao, Y.; Wang, B. Mapper Interactive: A scalable, extendable, and interactive toolbox for the visual exploration of high-dimensional data. In Proceedings of the 2021 IEEE 14th Pacific Visualization Symposium (PacificVis), Tianjin, China, 19–21 April 2021; pp. 101–110.
12. Brown, A.; Bobrowski, O.; Munch, E.; Wang, B. Probabilistic convergence and stability of random mapper graphs. *J. Appl. Comput. Topol.* **2021**, *5*, 99–140. [CrossRef]
13. Wasserman, L. Topological data analysis. *Annu. Rev. Stat. Its Appl.* **2018**, *5*, 501–532. [CrossRef]
14. Chazal, F.; Michel, B. An introduction to topological data analysis: Fundamental and practical aspects for data scientists. *Front. Artif. Intell.* **2021**, *4*, 667963. [CrossRef]
15. Lopez, J.E.; Datta, E.; Ballal, A.; Izu, L.T. Topological Data Analysis of Electronic Health Record Features Predicts Major Cardiovascular Outcomes After Revascularization for Acute Myocardial Infarction. *Circulation* **2022**, *146*, A14875. [CrossRef]
16. Yan, Y.; Ivanov, K.; Cen, J.; Liu, Q.H.; Wang, L. Persistence landscape based topological data analysis for personalized arrhythmia classification. *preprints* 2019.
17. Falsetti, L.; Rucco, M.; Proietti, M.; Viticchi, G.; Zaccone, V.; Scarponi, M.; Giovenali, L.; Moroncini, G.; Nitti, C.; Salvi, A. Risk prediction of clinical adverse outcomes with machine learning in a cohort of critically ill patients with atrial fibrillation. *Sci. Rep.* **2021**, *11*, 18925. [CrossRef]
18. Safarbali, B.; Hashemi Golpayegani, S.M.R. Nonlinear dynamic approaches to identify atrial fibrillation progression based on topological methods. *Biomed. Signal Process. Control* **2019**, *53*, 101563. [CrossRef]
19. Goldberger, A.L.; Amaral, L.A.; Glass, L.; Hausdorff, J.M.; Ivanov, P.C.; Mark, R.G.; Mietus, J.E.; Moody, G.B.; Peng, C.K.; Stanley, H.E. PhysioBank, PhysioToolkit, and PhysioNet: Components of a new research resource for complex physiologic signals. *Circulation* **2000**, *101*, e215–e220. [CrossRef]
20. Graff, G.; Graff, B.; Pilarczyk, P.; Jabłoński, G.; Gąsecki, D.; Narkiewicz, K. Persistent homology as a new method of the assessment of heart rate variability. *PLoS ONE* **2021**, *16*, e0253851. [CrossRef]
21. Ling, T.; Zhu, Z.; Zhang, Y.; Jiang, F. Early Ventricular Fibrillation Prediction Based on Topological Data Analysis of ECG Signal. *Appl. Sci.* **2022**, *12*, 10370. [CrossRef]

22. Mjahad, A.; Frances-Villora, J.V.; Bataller-Mompean, M.; Rosado-Muñoz, A. Ventricular Fibrillation and Tachycardia Detection Using Features Derived from Topological Data Analysis. *Appl. Sci.* **2022**, *12*, 7248. [CrossRef]
23. Caffrey, S.L.; Willoughby, P.J.; Pepe, P.E.; Becker, L.B. Public use of automated external defibrillators. *N. Engl. J. Med.* **2002**, *347*, 1242–1247. [CrossRef] [PubMed]
24. Delhomme, C.; Njeim, M.; Varlet, E.; Pechmajou, L.; Benameur, N.; Cassan, P.; Derkenne, C.; Jost, D.; Lamhaut, L.; Marijon, E.; et al. Automated external defibrillator use in out-of-hospital cardiac arrest: Current limitations and solutions. *Arch. Cardiovasc. Dis.* **2019**, *112*, 217–222. [CrossRef]
25. Kamp, N.J.; Al-Khatib, S.M. The subcutaneous implantable cardioverter-defibrillator in review. *Am. Heart J.* **2019**, *217*, 131–139. [CrossRef]
26. Friedman, P.; Murgatroyd, F.; Boersma, L.V.; Manlucu, J.; O'Donnell, D.; Knight, B.P.; Clémenty, N.; Leclercq, C.; Amin, A.; Merkely, B.P.; et al. Efficacy and safety of an extravascular implantable cardioverter–defibrillator. *N. Engl. J. Med.* **2022**, *387*, 1292–1302. [CrossRef]
27. Jiang, F.; Xu, B.; Zhu, Z.; Zhang, B. Topological Data Analysis Approach to Extract the Persistent Homology Features of Ballistocardiogram Signal in Unobstructive Atrial Fibrillation Detection. *IEEE Sens. J.* **2022**, *22*, 6920–6930. [CrossRef]
28. Ignacio, P.S.; Bulauan, J.A.; Manzanares, J.R. A Topology Informed Random Forest Classifier for ECG Classification. In Proceedings of the 2020 Computing in Cardiology, Rimini, Italy, 13–16 September 2020; pp. 1–4.
29. Ignacio, P.S.; Dunstan, C.; Escobar, E.; Trujillo, L.; Uminsky, D. Classification of single-lead electrocardiograms: TDA informed machine learning. In Proceedings of the 2019 18th IEEE International Conference on Machine Learning and Applications (ICMLA), Boca Raton, FL, USA, 16–19 December 2019; pp. 1241–1246.
30. Byers, M.; Hinkle, L.B.; Metsis, V. Topological Data Analysis of Time-Series as an Input Embedding for Deep Learning Models. In Proceedings of the IFIP International Conference on Artificial Intelligence Applications and Innovations, Crete, Greece, 17–20 June 2022; Springer: Cham, Switzerland, 2022; pp. 402–413.
31. Seversky, L.M.; Davis, S.; Berger, M. On time-series topological data analysis: New data and opportunities. In Proceedings of the IEEE Conference on Computer Vision and Pattern Recognition Workshops, Las Vegas, NV, USA, 27–30 June 2016; pp. 59–67.
32. Karan, A.; Kaygun, A. Time series classification via topological data analysis. *Expert Syst. Appl.* **2021**, *183*, 115326. [CrossRef]
33. Sun, F.; Ni, Y.; Luo, Y.; Sun, H. ECG Classification Based on Wasserstein Scalar Curvature. *Entropy* **2022**, *24*, 1450. [CrossRef]
34. Fraser, B.A.; Wachowiak, M.P.; Wachowiak-Smolíková, R. Time-delay lifts for physiological signal exploration: An application to ECG analysis. In Proceedings of the 2017 IEEE 30th Canadian Conference on Electrical and Computer Engineering (CCECE), Windsor, ON, Canada, 30 April–3 May 2017; pp. 1–4.
35. Dlugas, H. Electrocardiogram feature extraction and interval measurements using optimal representative cycles from persistent homology. *bioRxiv* **2022**. [CrossRef]
36. Nicponski, J.; Jung, J.H. Topological data analysis of vascular disease: A theoretical framework. *Front. Appl. Math. Stat.* **2020**, *6*, 34. [CrossRef]
37. Bresten, C.L.; Kweon, J.; Chen, X.; Kim, Y.H.; Jung, J.H. Preprocessing of general stenotic vascular flow data for topological data analysis. *bioRxiv* **2021**. [CrossRef]
38. Tokodi, M.; Shrestha, S.; Ashraf, M.; Casaclang-Verzosa, G.; Sengupta, P. Topological Data Analysis for quantifying inter-patient similarities in cardiac function. *J. Am. Coll. Cardiol.* **2019**, *73*, 751. [CrossRef]
39. Tokodi, M.; Shrestha, S.; Bianco, C.; Kagiyama, N.; Casaclang-Verzosa, G.; Narula, J.; Sengupta, P.P. Interpatient similarities in cardiac function: A platform for personalized cardiovascular medicine. *Cardiovasc. Imaging* **2020**, *13*, 1119–1132.
40. Fasy, B.T.; Lecci, F.; Rinaldo, A.; Wasserman, L.; Balakrishnan, S.; Singh, A. Confidence sets for persistence diagrams. *Ann. Stat.* **2014**, *42*, 2301–2339. [CrossRef]
41. Chazal, F.; Fasy, B.T.; Lecci, F.; Rinaldo, A.; Wasserman, L. Stochastic convergence of persistence landscapes and silhouettes. In Proceedings of the Thirtieth Annual Symposium on Computational Geometry, Kyoto, Japan, 8–11 June 2014; pp. 474–483.
42. Wadhwa, R.R.; Williamson, D.F.; Dhawan, A.; Scott, J.G. TDAstats: R pipeline for computing persistent homology in topological data analysis. *J. Open Source Softw.* **2018**, *3*, 860. [CrossRef]
43. Bauer, U.; Kerber, M.; Reininghaus, J.; Wagner, H. Phat–persistent homology algorithms toolbox. *J. Symb. Comput.* **2017**, *78*, 76–90. [CrossRef]
44. Bauer, U.; Kerber, M.; Reininghaus, J. Distributed computation of persistent homology. In Proceedings of the 2014 Sixteenth Workshop on Algorithm Engineering and Experiments (ALENEX), Portland, OR, USA, 5 January 2014; pp. 31–38.
45. Zhang, S.; Xiao, M.; Wang, H. GPU-accelerated computation of Vietoris-Rips persistence barcodes. *arXiv* **2020**, arXiv:2003.07989.
46. Kerber, M.; Nigmetov, A. Efficient approximation of the matching distance for 2-parameter persistence. *arXiv* **2019**, arXiv:1912.05826.

entropy

MDPI

Article

Predictive Modeling of Heart Rate from Respiratory Signals at Rest in Young Healthy Humans

Carlos M. Gómez [1,*], Vanesa Muñoz [1] and Manuel Muñoz-Caracuel [1,2]

1 Human Psychobiology Laboratory, Experimental Psychology Department, University of Seville, 41018 Seville, Spain; lmunnoz@us.es (V.M.); manuelmunozcaracuel@outlook.com (M.M.-C.)

2 Hospital Universitario Virgen del Rocio, 41013 Seville, Spain

* Correspondence: cgomez@us.es

Abstract: Biological signals such as respiration (RSP) and heart rate (HR) are oscillatory and physiologically coupled, maintaining homeostasis through regulatory mechanisms. This report models the dynamic relationship between RSP and HR in 45 healthy volunteers at rest. Cross-correlation between RSP and HR was computed, along with regression analysis to predict HR from RSP and its first-order time derivative in continuous signals. A simulation model tested the possibility of replicating the RSP–HR relationship. Cross-correlation results showed a time lag in the sub-second range of these signals (849.21 ms $\pm$ SD 344.84). The possible modulation of HR by RSP was mediated by the RSP amplitude and its first-order time derivative (in 45 of 45 cases). A simulation of this process allowed us to replicate the physiological relationship between RSP and HR. These results provide support for understanding the dynamic interactions in cardiorespiratory coupling at rest, showing a short time lag between RSP and HR and a modulation of the HR signal by the first-order time derivative of the RSP. This dynamic would optionally be incorporated into dynamic models of resting cardiopulmonary coupling and suggests a mechanism for optimizing respiration in the alveolar system by promoting synchrony between the gases and hemoglobin in the alveolar pulmonary system.

Keywords: heart rate; respiration; predictive modeling; human physiology; simulation

1. Introduction

Metabolic and autonomic processes in humans and animals are regulated by control systems that facilitate adaptation and survival in response to internal or external changes. This adaptive process is essential for maintaining stability and is a well-documented phenomenon known as homeostasis, which plays a critical role in maintaining an efficient functional state [1]. Canon [2] introduced the term to describe the feedback operation in physiological systems and the maintenance of equilibrium. It is now recognized as an inherent self-regulatory system within biological systems that is necessary for the state of normal functioning [3].

Biological signals of the nervous system typically exhibit inherent oscillations with varying characteristics, including frequency and temporal variations. The precise variability of these signals is often unclear [4]. Oscillations are influenced by environmental factors and intrinsic processes, including homeostasis and feedback. For instance, spontaneous oscillations in heart rate (HR) have been extensively studied by frequency decomposition, revealing their associations with various activities of the sympathetic and parasympathetic nervous systems [5]. Similar studies have been extended to other biological systems.

HR and respiration signals (RSP) and their oscillations are amplitude-modulated and synchronized [6,7], considering that respiration influences the cardiac response. Inspiration increases HR, while expiration decreases it, a phenomenon known as respiratory sinus arrhythmia (RSA), which is widely accepted as an index of vagal activity enhancing pulmonary gas exchange [7,8]. The physiological system primarily responsible for the interaction between respiration and circulation is the autonomic nervous system (ANS). However, the physiological mechanisms underlying the modulation of cardiac responses by respiration are complex and multifactorial. Both HR and RSP signals are influenced by external and internal mechanisms. Cardiovascular variability is mainly modulated by the two branches of the ANS, which affect the redistribution of blood volume, blood flow, heart rate, and heart variability, but its modulation is also proposed to be largely influenced by breathing. These effects are complex and affect cardiac responses at multiple levels.

The relationship between RSP and HR, often referred to as cardiorespiratory coupling (CRC), has been successfully modeled by complexity measures [9–12]. These studies focusing on CRC and nonlinear dynamics provide insights into physiological regulation in the context of adaptive mechanisms to stress, aging, and physical training. For instance, stressors such as dehydration have been related to reduced synchronization indices and altered directional relationships between cardiovascular and respiratory variables [9]. Aging further modulates these dynamics, with the resting state showing weakened respiratory contributions to cardiac regulation, while stress conditions amplify these impairments [10]. In contrast, athletes have shown stronger CRC compared to sedentary individuals, reflecting enhanced autonomic synchronization [12]. Specific training interventions, such as inspiratory muscle training, have shown the potential to enhance CRC by promoting central respiratory and autonomic adaptations, with effects observed during active challenges and subtle improvements at rest [11].

The use of linear relationships involving the RSP and its first-time derivative, as a possible causal signal for HR has been explored [13], suggesting an important role for stretch receptors in modulating HR. However, the continuous dynamic relationship between the RSP and HR signals has not been intensively explored. The demonstration of the role of the first-order time derivative of the RSP term in the prediction of HR from RSP would suggest the possibility of linearizing this relationship in the whole RSP cycle during resting state respiration and, given the quasi-sinusoidal character of the RSP signal, it would be possible to model the regulation of the time-lag between RSP and HR by controlling the gains of the RSP and its first-time derivative. Such a model, once empirically validated, would allow to explore by simulation of the HR signal, the regulation of the physiological delay between the RSP and the HR signals, by modulating the amplitude of the signals conveyed to the central nervous system by slow and fast adapting pulmonary stretch receptors [13].

The respiratory rhythm is typically generated by respiratory peacemaker cells in the brainstem, and its oscillations are modulated by central and peripheral reflexes that adapt in response to environmental demands [14]. The origin of the RSA and/or its modulation is currently a matter of debate, several hypotheses could be complementary. In fact, such a complex regulation of the cardiorespiratory coupling would give the system great robustness to maintain an adequate level of function. These proposed mechanisms for the generation and/or regulation of RSA are (i) a central mechanism exerted by the brainstem sympathetic and parasympathetic nuclei, which would be controlled by the central respiratory nuclei through a central pacemaker [15–18]; (ii) one of the main regulatory nodes of RSA would be in the baroreceptor reflex, as evidence from baroreceptor suction shows its strong influence on RSA, for instance, in forced apnea, RSA is induced by cyclic suction stimulating carotid baroreceptors, suggesting a crucial role for the baroreceptor reflex. Under more natural conditions, the inspiratory phase would increase venous return and then increase arterial pressure, inducing an increase in baroreceptor activity that would activate vagal nuclei in the brainstem [14]; (iii) the activation of pulmonary stretch receptors, generating the Hering-Breuer reflex, which by modifying the central pattern generator of respiration would also modify the vagal response [19]; (iv) but also mechanical pressure exerted over the atrial sinus can modulate the RSA, the so-called Bainbridge reflex [20], and (v) hypercapnia, oxygen pressure, and acidosis acting on central and peripheral chemoreceptors can modulate respiratory sinus arrhythmia [21–23]. The controversy regarding the relative importance of central or reflexive mechanisms in the generation and/or modulation of RSA is a long-standing debate with a difficult answer [24] and will not be directly addressed in this report. It is worth noting that both HR and RSP serve the primary purpose of maintaining homeostasis through the balanced and proper functioning of these physiological mechanisms.

In addition to RSA, other types of cardiorespiratory coupling include cardiorespiratory phase synchronization (CRS) and cardiorespiratory coordination (CRCo) [25,26]. CRS refers to the tendency of the heartbeat to align with a specific phase of the respiratory cycle, often observed during deep sleep and under anesthesia. On the other hand, CRCo examines the relationship between the heartbeat and the preceding and following respiratory onset in the time domain. While CRS allows us to analyze how the timing of heartbeats aligns with specific phases of the respiratory cycle [9], CRCo can adapt to variations in respiratory cycle lengths, capturing dynamic changes in the heart–respiration relationship, and some studies have shown their differences in the degree of information provided by the two approaches (relaxation for CRS and stress for CRCo), the CRCo being independent of the length of the respiratory cycle unlike the CRS [26]. These terms could be confused and related to the RSA, but their effects can be observed both independently and in conjunction with RSA. Although there are relevant terms in the understanding of cardiorespiratory coupling, the present study does not attempt to integrate them as they are beyond the main objective of the study.

The present report aims to improve the understanding of the predictive heart rate estimation based on respiratory signals. By analyzing this relationship from a statistical and dynamic point of view, this will provide more accurate information for the modeling of this process. Specifically, we explore the possibility of predicting the continuous heart rate cardiac signal from the respiratory, inspiratory, and expiratory signals in healthy young adults during a spontaneous resting state. Furthermore, the present results are intended to highlight some dynamic aspects of the relationships between RSP and cardiac signals, which would be optionally useful for physiologists and modelers to incorporate in functional descriptions of the relationship between RSP and HR in a resting state.

The main objective of the present report is to model the continuous linear dynamic relationship between the RSP and HR signals, incorporating the first-order time derivative of the RSP [13]. This will support the existence of an anticipatory signal of the respiratory state of the lungs (inspiratory inflation and expiratory deflation) which would help to modulate the time-lag between RSP and HR. Obtaining such a relationship would suggest a possible physiological mechanism to optimize gas exchange by better aligning the RSP and HR. This could possibly be supported by pulmonary fast and slow adapting stretch receptors and/or baroreceptors and chemoreceptors, which would modulate the vagal response and then HR. It can be hypothesized that the continuous HR signal can be predicted from the continuous RSP signal and its first-time derivative, and as a consequence, modulating the gain of these terms will allow us to adjust the time-lag between the respiratory and cardiac signals.

2. Materials and Methods

2.1. Participants

A sample of 45 volunteer subjects (18 males and 27 females, mean age = 26.71 $\pm$ 4.41 SD) participated in the present study. All of them were healthy subjects without any neurological or psychological disorders. They were university students who reported no cardiopulmonary problems or any other major health issues. They were recruited through advertisements at the University of Seville. Before the study, they were informed about the procedure and then signed an informed consent form. The study was approved by the Bioethical Committee of the Junta de Andalucía (1045-N-21 and 1717-N-22) and followed the rules of the latest revision of the Declaration of Helsinki for human research.

2.2. Experimental Session

Peripheral ECG and RSP signals were recorded using an MP160 (BIOPAC Systems, Goleta, CA, USA) with two amplifier modules (ECG-100C and RSP-100C) (Figure 1). The ECG was collected using three Ag-AgCl lead electrodes (positive, negative, and ground, TSD203 transducer) placed on the left wrist, right wrist, and right ankle, respectively. The RSP signal was recorded by a transducer band placed on the chest (TSD201 transducer). Participants were instructed to move as little as possible and to look at the screen for 3 min without any task or response. Respiration during this period was unassisted and spontaneous. The gain was set to 1000 in the ECG and 10 in the RSP amplifier. The sampling rate was 1000 Hz. Data acquisition was performed with AcqKnowledge v.5.0.1 software (BIOPAC Systems).

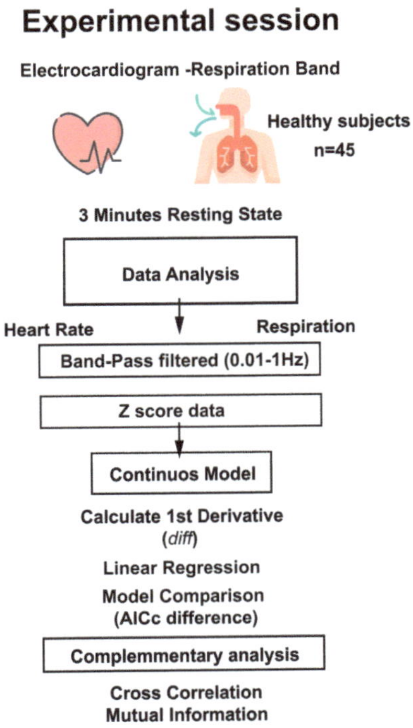

Figure 1. Flowchart of signal processing. Flowchart showing the experimental session and subsequent data analysis. AICc: Akaike information criterion.

2.3. Data Analysis

For data analysis (Figure 1), the ECG signal was first processed in AcqKnowledge software, to calculate the HR time series from the ECG signal. The processed signals were exported to Matlab R2019b (MathWorks), where a custom script was used to extract 3 min of spontaneous activity for each subject (Figure 2). To ensure that the analysis captured relevant physiological oscillations while minimizing noise, both HR and RSP signals were band-pass filtered between 0.1 Hz and 1 Hz using a zero-phase filter filtfilt. This range was chosen to include the RSA dynamic, considering that the respiratory rate is around 0.1 to 0.3 Hz, and the HR variability ranges are around 0.04 to 0.4 Hz, and specifically, the RSA effect has been related to the HF component > 0.15 Hz [27]. Neither the RSP nor the HR signal showed clear trends after applying the high-pass filter at 0.01 Hz. The filtered signals were then standardized using z-score.

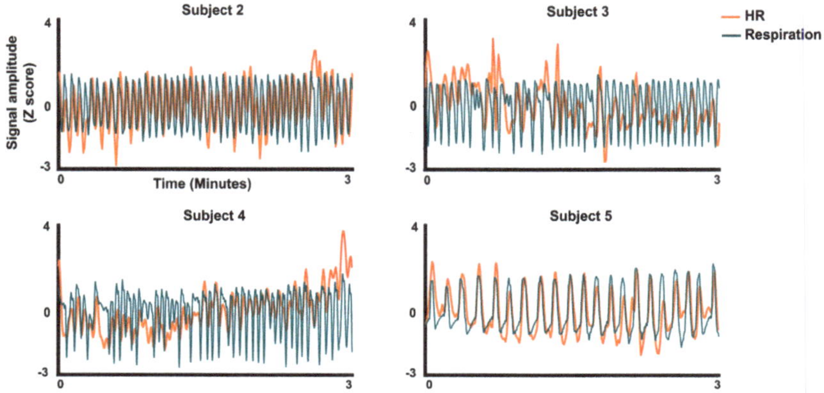

Figure 2. Signal plots. Three minutes recording for the HR and respiratory signals in z-score values for four subjects.

2.4. Signal Analysis

To analyze the possible time-lag delay of the HR signal with respect to the RSP signal, a cross-correlation analysis between the two signals was calculated in Matlab using the xcorr function. In addition to cross-correlation, mutual information (MI) analysis was performed to assess the non-linear dependencies between the HR and RSP signals. Mutual information quantifies the amount of information that one random variable contains about another random variable [28] and captures both linear and non-linear relationships. The MI was calculated using the joint probability distribution of the signals, derived from their joint histogram, and the marginal probabilities for each signal. As mentioned above, these analyses were performed on the z-score signals to ensure consistent results.

To test for a possible dependence of the continuous HR signal on the continuous RSP signal and its first time derivative, the following continuous models were tested by regression analysis in each subject on the continuous RSP signals and (as shown in Figures 2 and 3) using the Matlab function fitlm.

Continuous model 1 (c-model1):

$$\mathrm{HRm(t)} = \mathrm{b0} + \mathrm{b1} * \mathrm{RSP(t)} \quad (1)$$

Continuous model 2 (c-model2) (based on Kapidžić et al., 2014 [13]):

$$\mathrm{HRm(t)} = \mathrm{b0} + \mathrm{b1} * \mathrm{RSP(t)} + \mathrm{b2} * \mathrm{dRSP/dt} \quad (2)$$

HRm(t): Modelled continuous heart rate signal from the respiratory signal.
RSP (t): Continuous respiratory signal.
dRSP/dt: First-time derivative of RSP(t).

Then, the best model was calculated by the Akaike Information Criterion (AIC) subject by subject [29]. The operationalized variables were the empirical variables RSP and HR (both in z-score), and the computed variables HRm(t), and the first-time derivative of RSP (dRSP/dt). The time-lag values of RSP and HR were obtained by cross-correlation. Statistical analyses were performed in Matlab R2019b (MathWorks) and the Statistical Package for the Social Sciences 26 (SPSS). A simulation, described in the next section, was built up to show that the computation of the RSP time derivative would be able to modulate the time lag between RSP and HR.

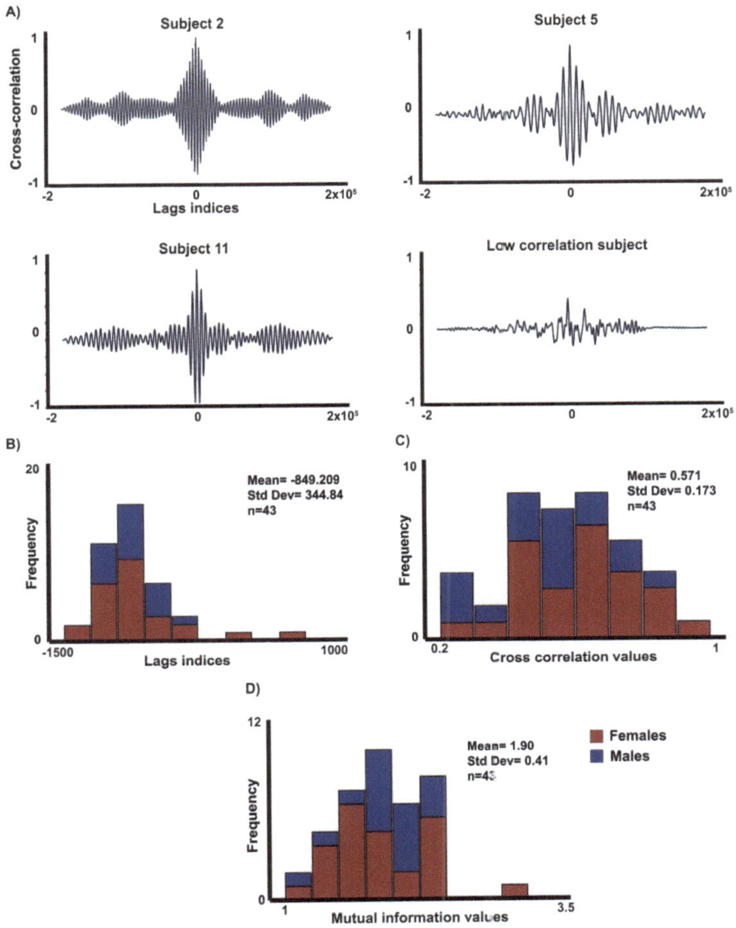

Figure 3. Cross-correlation and mutual information analysis. (**A**) Cross-correlation analysis between the RSP and HR signals in three subjects with the highest correlation coefficients (0.817–0.93) and one subject with low correlation (0.218). (**B**) Histogram with the time-lags indices excluding the subjects with higher time-lag values and very poor cross-correlation. (**C**) Histogram of the cross-correlation values without excluding subjects. (**D**) Histogram with the mutual information values without excluding subjects.

3. Results

First, we were interested in how the RSP and HR cycles would be related to a possible delay between the RSP and the HR cycles, to be determined by cross-correlation between these two signals. For the cross-correlation analysis (Figure 3), the general results showed negative time-lag indices and high peak correlation when the RSP signal was cross-correlated with HR. There were no significant gender differences in the time-lag and cross-correlation analysis. However, in two subjects the lag indices were extremely large and showed a low correlation peak. These two outliers were detected by the isoutlier function. Excluding these two subjects, the mean value of the lag indices was 849.21 $\pm$ SD 344.84, indicating a delayed response for the HR compared to the RSP signal. Similarly, taking into account the relevance of the non-linearities in the study of the influence of the RSP in HR, an MI analysis was added, finding mean values of 1.90 $\pm$ SD 0.41. The histogram shows that the MI values are clustered around a moderate level of shared information between HR and RSP signals. No significant effect of gender was found for this metric. To ensure that the linear approach captures the most relevant information about the relationship of the RSP and HR at rest, a correlation analysis was calculated between these two metrics, which showed a significant correlation between MI and cross-correlation ($R = 0.498$; $p < 0.001$). These results suggest that at rest, most of the common information between HR and RSP signals is captured by linear dependencies.

To test the dynamic relationship between RSP and HR signals, the continuous models described in Section 2 were applied and tested (c-model1 and c-model2). Note that c-model1 is similar to relating the empirical HR to the empirical RSP, as HRm in c-model1 is linearly related to RSP. Figure 4A shows a pattern similar to a closed trajectory for the relationship of RSP to HR. This closed trajectory is suggested in Figure 4B to be due to the time-lag between RSP and HR. However, when the predicted signal HRm generated by the c-model2 is compared with HR, the phase-lag with HR disappears (Figure 4B orange lines), and a trend showing the linearization of the relationship between the predicted HRm values of c-model2 and empirical HR appears (Figures 4C and 5). The results for all subjects are shown in Supplementary Table S1 and Supplementary Figures S1 and S2. The most explanatory model according to the AICc criterion is c-model2 when compared to c-model1 (45/45 cases). The slope of the first term (proportional to the RSP signal) shows a preponderance of positive slopes (32/45), while the second term (proportional to the first-time derivative of the RSP) shows a preponderance of negative slopes: 44/45. When the total sample of subjects is plotted for the linear regressions between HR vs. RSP, and HR vs. predicted values from c-model2 (HR vs. HRm), the slope of the c-model2 (HRm vs. HR) was always positive and steeper than in c-model1 (HR vs. RSP) (Supplementary Figures S1 and S2).

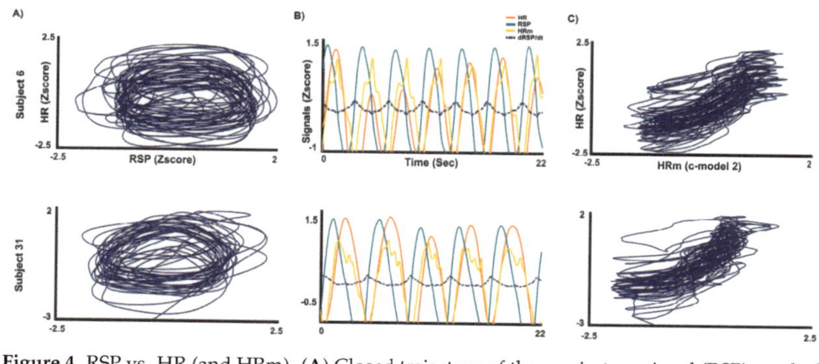

Figure 4. RSP vs. HR (and HRm). (**A**) Closed trajectory of the respiratory signal (RSP) vs. the heart rate signal (HR). (**B**) This figure shows the trend to linearization between the predicted HRm and HR signal when the c-model2, which includes a term of the first-time derivative of RSP, is applied. Please observe the high overlapping between the predicted HRm values from cmodel-2 and the HR signal (in orange). (**C**) When the predicted values of c-model2 (HRm) are represented vs. HR, a linearity of the relationship emerges.

These results suggest that the consideration of dRSP/dt improves the prediction of the HR signal. A possible model for the relationship between peripheral receptor activity and the HR response is shown in Figure 6. To test the hypothesis that the derivative term would have the effect of reducing the delay between RSP and HR, the following simulation (script in the Supplementary Material) is sketched in the diagram of Figure 6. First, (i) a simulated RSP signal (RSP(t)) is generated from an asymmetric and smoothed triangular signal (Simulated_RSP variable in the script's code) (Figure 7A), (ii) and then a delayed HRm signal is obtained by inserting a delay time into the RSP (t) (delayed_HR in the code), (iii) then dRSP/dt (deriv_RSP in the code) is calculated, which allows (iv) the HRm variable (advanced_HR in the code) to be obtained as a direct application of c-model2 by summing the dRSP/dt term modulated by gain2 to the RSP(t). HRm (advanced_HR) would simulate the empirically recorded HR. By changing the gain2 of the dRSP/dt, it is possible to adjust the delay between the RSP(t) signal (Simulated_RSP in the code) and the HRm (advanced_HR in the code) (Figure 7B). Alternatively, the time-lag can be modulated by changing the gain1 of the RSP(t). Not only the time-lag but also the amplitude of HRm can be modulated by a multiplicative gain3 factor. The dashed line in Figure 6 suggests the possibility, not included in the model, that chemoreceptors would modulate the values of gain1 and/or gain2 values to optimize respiratory gas exchange at the pulmonary alveolar system.

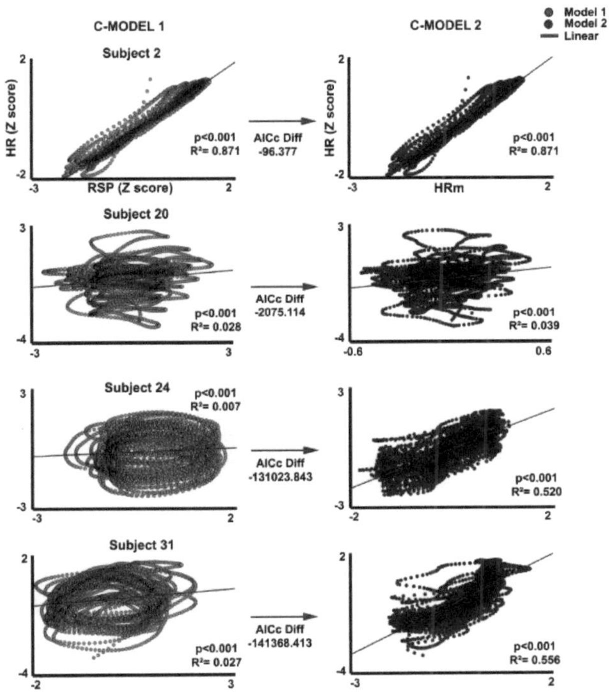

Figure 5. Linear regression of representative subjects. Graphical representation of the linear regressions of HR vs. RSP (left side) and HR vs. the predicted values of c-model2 (HRm). Two subjects with a low AICc difference and two subjects with a high AICc difference are displayed. The same analysis for the complete sample (45 subjects) is displayed in Supplementary Figures S1 and S2. AICc: Akaike information criterion corrected.

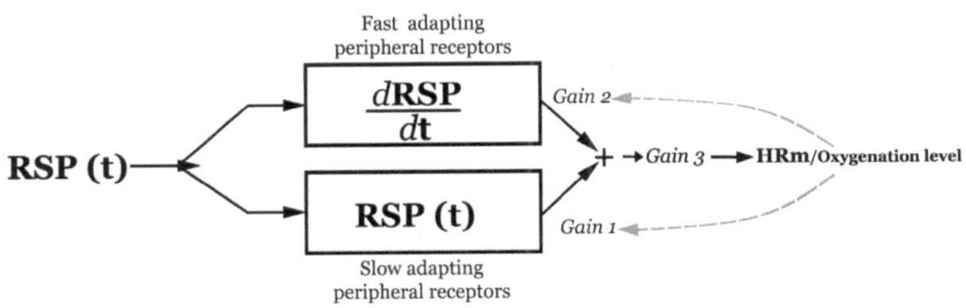

Figure 6. Model of the relationship between activity in the peripheral receptors and the heart rate response, following cmodel-2. RSP: continuous respiratory signal (Recorded_RSP in the script code), HRm: modelled heart rate (Advanced_HR in the script code).

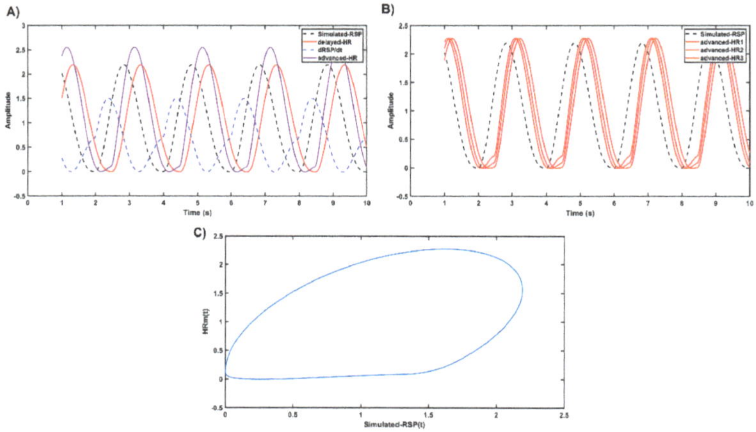

Figure 7. Simulation of heart rate from the respiration signal. (**A**) The simulated RSP(t), its time derivative, and delayed and advanced simulated HRm by c-model2 are displayed. (**B**) The modulations of the time-lags between the simulated RSP(t) and HRm for three different values of the gain2 are displayed. (**C**) Closed trajectory between the simulated RSP(t) and HRm replicating the relationship between the experimentally recorded RSP and HR.

Note that the delay between RSP(t) and HRm included in the model is a consequence of the whole process. By modulating the gain1 for the RSP(t) term, gain2 for the dRSP/dt, or both, it is possible to balance the time-lag between RSP(t) and HRm. The script in the Supplementary Material allows to demonstrate how the adjustment of the gain 1 and/or 2 allows to modulate the time-lag between RSP and HR.

Figure 7A shows the computation of the simulated HRm as a delayed linear relationship of the RSP(t) signal. By adding the derivative of respiration (dRSP/dt) to the RSP(t), the delay of the heart rate with respect to the RSP(t) can be reduced, resulting in the advanced HRm, which is intended to simulate the empirically recorded HR. By changing the gain2 of the dRSP/dt term, the time-lag between the RSP(t) and HRm signals can be modulated (Figure 7B). The amplitude of the HRm signal can also be modulated (gain3). Finally, the closed trajectory observed in most of the subjects for the RSP vs. HR signals, as shown in Figure 4A and Supplementary Figures S1 and S2, is obtained by plotting the simulated RSP(t) vs. the computed HRm (Figure 7C). The physiological feasibility and possible adaptive value of the model will be discussed. Similar results as in Figure 7A–C can be obtained by modulating gain1 (see script in the Supplementary Material).

4. Discussion

The dynamic relationship between the RSP signal and the HR signal is supported by the significant cross-correlation, mutual information, and time-lag delay between these two signals. The c-model2 suggests that the amplitude of the heartbeat frequency during RSA would be related to the amplitude of the RSP signal and its first-order time derivative of the RSP amplitude. The inclusion of the time derivative of the RSP allows linearization of the modeling of the cardiopulmonary relationship. This would allow anticipating the state of the lung (inflation and deflation), reducing the delay between the activation of the stretch receptors and the cardiac response, and facilitating gas exchange in the alveolar system, as previously suggested [22,30]. The proposed model has been simulated and allows to confirm that the modulation of the gain of the RSP(t) and/or the dRSP/dt term allows to regulate the time-lag between RSP and HR.

The inclusion of MI analysis in the present report provided additional insights into the nonlinear dynamics of the dynamic relationship between HR and RSP at rest. The results showed values around a moderate level of shared information between HR and RSP signals, similar to the cross-correlation results, and supporting the existing literature on their interconnection [6,12,16,22,31]. Furthermore, the significant correlation between MI and cross-correlation suggests that linear dependencies account for a substantial proportion of the shared information between HR and RSP signals under spontaneous resting state conditions. This finding reinforces the utility of linear approaches in capturing the dynamics of both signals at rest while highlighting the complementary role of non-linear metrics such as MI in exploring additional features in the relationship between these physiological signals. While non-linear analyses provide valuable complementary information, the strong performance of linear metrics suggests that much of the shared information between these signals could be attributed to stable, linear interactions when analyzing young healthy subjects under spontaneous resting state conditions. Possibly, nonlinearities would be more relevant for explaining RSP–HR interactions under stressful, pathological, and aging conditions.

The cross-correlation results show that the RSP signal precedes the HR, the latter with a mean delay of 849.21 ms, with oscillations that are coupled in most subjects, according to the concept of the RSA of HR causally linked to RSP. These results support the well-known hypothesis that respiratory cycles influence the cardiac response [16,31]. The observed time-lag values are consistent with some previous measurements, for instance, Saul et al. [32] proposed that R–R interval changes follow breathing with a time lag of about 0.3 s. The presence of a delay between the respiratory and cardiac cycles [33] does not necessarily resolve the controversy about a central or peripheral mechanism for generating RSA [24], since different modulatory pathways with synchronized but different time delays would explain the respiratory-to-cardiac cycle time-lag. This result is consistent with proposals on the controlling role of respiration over cardiac cycles by central [34]; or peripheral mechanisms linked to respiratory [19], or respiratory-related signals [14,21,22]. Therefore, a time-lag delay between the respiratory and cardiac cycles should be included in the dynamic modeling of RSA.

The relationship between both signals is more evident when considering the maximally explanatory model for HR from RSP (c-model2), in which the inclusion of the first-order time derivative of the respiratory signal improves the ability of RSP to predict HR. A result, previously obtained for the inspiratory phase, and incorporating the RSP and its time derivative, is used to predict HR [13]. From a modeling perspective, the first-order time derivative allows a transition from a closed trajectory for the relationship between RSP and. HR to a more linear one. The dependence of the RSP signal on HR has previously been observed and quantified using autoregressive moving average techniques [31], demonstrating the influence of the time evolution of RSP (and arterial pressure) on HR. However, such modeling does not establish a precise relationship between the time derivative of RSP and HR, a relationship that has been previously described [13], suggesting possible mechanisms for pathological conditions. A potential advantage of the c-model2 is that it suggests the presence of an anticipatory computation of the respiratory phase by partially discounting the time delay between pulmonary stretch receptors activation and HR response. This could optimize gas exchange [22,30] in the alveolar pulmonary system between RSP and HR by a precise time-lag between both signals.

An important point to note about c-model2 is that when the predicted values of HR from c-model2 (HRm) are regressed with HR, a linearization with a positive slope appears. This result should be interpreted from a modeling perspective as if RSP, once compensated by the rate of change (second term of c-model2), linearly drives the cardiac cycle during the resting state, supporting the view of a modulation of HR from the respiratory cycle [7,8]. One suggestion would be that slowly adapting stretch receptors would carry the first term of c-model2 (proportional to the RSP signal) and the fast-adapting receptors, which are more sensitive to changes in RSP, would carry the second term of c-model2, proportional to the time derivative of RSP. The presence of rapid and fast adapting stretch receptors in the respiratory muscles would carry the RSP and dRSP/dt signals to central vagal centers [19] without discarding the possible role of arterial pressure mechanoreceptors in the carotid body. The possible role of stretch receptors must be considered tentatively, but would explain the influence of the RSP time derivative on HR. Of course, the cardiac response has much broader peripheral and central regulatory points, including the carotid body receptors, which are difficult to assess from the present results. However, the reported simulation must be understood in terms of signal processing modeling rather than as a specific physiological proposal that would require pharmacological or physiological intervention and/or the recording of peripheral and central nervous system generators.

The possibility of obtaining a precise time-lag between RSP and HR during RSA, by regulating the gain of RSP or dRSP/dt, is important to optimize the synchrony between the inspiration phase at the alveolar level and the arrival of an optimal amount of deoxygenated blood at the alveolar level. This is a dynamically complex problem to be computed due to the physics of the gases and fluids passing through the bronchi and bronchioles and the different ramifications at the cardiovascular level and blood flow through them. The present model and simulation allow the possibility of regulating the time-lag by modifying the proportion of RSP and dRSP/dt signal at the neural level. However, modeling neural processing is also complex. To model the exact relationship between the contribution of the RSP and dRSP/dt to HR would require a lot of physiological information, taking into account the mechanics of the respiratory system to activate the stretch receptors, the activation of the receptors, the central processing, the activation of the heart, and the heart electrical transmission. Such an amount of physiological information would be impracticable in humans. However, the proposed model opens the possibility of finding the most adaptive and energetically efficient oxygenation level by regulating the cardiorespiratory time-lag during RSA, with chemoreceptors providing information about the efficacy of gas exchange [21–23]. Assessment of blood oxygenation levels would hypothetically induce changes in the model gain1 to optimize the synchrony between atmospheric gas pressure and hemoglobin concentration at the alveolus to optimize gas exchange [22,30]. But also, the possibility of modulating the delay between RSP and HR could be modulated by precise timing initiation of the inspiratory phase from the cardiac activity [35].

The present report makes some suggestions about the possible control of HR by peripheral stretch receptors related to the RSP cycle. However, to confirm this approach, the forced control of the RSP cycle, pharmacological approaches, and/or invasive recordings of the peripheral and central nervous system would be required to follow the neural feedback chain, central generators, and autonomic cardiac control. Also, testing in pathological conditions where the interaction between RSP and HR is disturbed would be very useful to understand this relationship.

5. Conclusions

In conclusion, to model the possible inter-relationship between the RSP and HR signals, a time-lag delay between RSP and HR, which could be reduced by taking into account the time derivative of the RSP signal, would give a good account of the RSP–HR cardiorespiratory dynamics during the resting state in young healthy subjects. The proposed model allows the linearization of the relationship between RSP and HR and, by modulating the gains of the RSP and/or dRSP/dt terms, allows the adjustment of the time-lag between RSP and HR, possibly allowing the optimization of the gas exchange between the vascular and pulmonary systems.

Supplementary Materials: The following supporting information can be downloaded at: https://www.mdpi.com/article/10.3390/e26121083/s1. Table S1. Linear model regression of continuous waves of HR and RSP z-score signals including the time derivatives of RSP. Figure S1. Linear regression of the continuous signal. Graphical representation of the linear regressions of HR vs. RSP (left side) and HR vs. the predicted values of c-model2 for the first 24 Subjects (1–24 Subjects). Figure S2. Linear regression of the continuous signal. Graphical representation of the linear regressions of HR vs. RSP (left side) and HR vs. the predicted values of c-model2 for the last 21 Subjects (25–45 Subjects). Supplementary code. Matlab custom script to perform a simulation that allows to replicate the physiological relationship between RSP and HR.

Author Contributions: Conceptualization, C.M.G.; methodology, C.M.G.; software, M.M.-C.; formal analysis, C.M.G. and V.M.; investigation, M.M.-C. and V.M.; data curation, C.M.G. and V.M.; writing—original draft preparation, C.M.G. and V.M.; writing—review and editing, C.M.G. and V.M.; visualization, V.M.; supervision, C.M.G.; funding acquisition, C.M.G. All authors have read and agreed to the published version of the manuscript.

Funding: This study was supported by Spanish Agencia Estatal de Investigación (AEI) (PID2022-139151OB-I00), and Vanesa Muñoz was funded by a scholarship from the Universidad de Sevilla (VIPPIT-2020-IV.3).

Institutional Review Board Statement: The experiment followed the rules of the latest revision of the Declaration of Helsinki for Human Research (2013) and was approved by the Ethics Committee of Junta de Andalucía (1717-N-22).

Data Availability Statement: The data presented in this study are openly available on FigShare at https://doi.org/10.6084/m9.figshare.27332226.v1.

Acknowledgments: We would like to thank all the participants in the study for their patience and willingness to volunteer.

Conflicts of Interest: The authors declare no conflicts of interest.

References

1. Plaxton, W. Principles of metabolic control. In *Functional Metabolism: Regulation and Adaptation*; Storey, K.B., Ed.; John Wiley & Sons: Hoboken, NJ, USA, 2004; pp. 1–24.
2. Cannon, W.B. *The Wisdom of the Body*; W.W. Norton & Company: New York, NY, USA, 1932.
3. Zholtkevych, G.N.; Nosov, K.V.; Bespalov, Y.G.; Rak, L.I.; Abhishek, M.; Vysotskaya, E.V. Descriptive Modeling of the Dynamical Systems and Determination of Feedback Homeostasis at Different Levels of Life Organization. *Acta Biotheor.* **2018**, *66*, 3–199. [CrossRef] [PubMed]
4. McClintock, C.; Stefanovska, A. Chapter 1: Introduction. In *Physics of Biological Oscillators: New Insights into Non-Equilibrium and Non-Autonomous Systems*; Stefanovska, A., McClintock, C., Eds.; Springer: Cham, Switzerland, 2021.
5. Nitzan, M.; de Boer, H.; Turivnenko, S.; Babchenko, A.; Sapoznikov, D. Power spectrum analysis of spontaneous fluctuations in the photoplethysmographic signal. *J. Basic Clin. Physiol. Pharmacol.* **1994**, *5*, 269–276. [CrossRef] [PubMed]
6. Schäfer, C.; Rosenblum, M.G.; Kurths, J.; Abel, H.H. Heartbeat synchronized with ventilation. *Nature* **1998**, *392*, 239–240. [CrossRef] [PubMed]
7. Cairo, B.; de Abreu, R.M.; Bari, V.; Gelpi, F.; De Maria, B.; Rehder-Santos, P.; Sakaguchi, C.A.; da Silva, C.D.; De Favari Signini, É.; Catai, A.M.; et al. Optimizing phase variability threshold for automated synchrogram analysis of cardiorespiratory interactions in amateur cyclists. *Philos. Trans. A Math. Phys. Eng. Sci.* **2021**, *379*, 2212. [CrossRef]
8. Yasuma, F.; Hayano, J. Augmentation of respiratory sinus arrhythmia in response to progressive hypercapnia in conscious dogs. *Am. J. Physiol. Heart Circ. Physiol.* **2001**, *280*, H2336–H2341. [CrossRef] [PubMed]
9. Zhang, Q.; Patwardhan, A.R.; Knapp, C.F.; Evans, J.M. Cardiovascular and cardiorespiratory phase synchronization in normovolemic and hypovolemic humans. *Eur. J. Appl. Physiol.* **2015**, *115*, 417–427. [CrossRef] [PubMed]
10. Porta, A.; Faes, L.; Bari, V.; Marchi, A.; Bassani, T.; Nollo, G.; Perseguini, N.M.; Milan, J.; Minatel, V.; Borghi-Silva, A.; et al. Effect of age on complexity and causality of the cardiovascular control: Comparison between model-based and model-free approaches. *PLoS ONE* **2014**, *9*, e89463. [CrossRef]
11. de Abreu, R.M.; Catai, A.M.; Cairo, B.; Rehder-Santos, P.; Silva, C.D.; Signini, É.D.; Sakaguchi, C.A.; Porta, A. A Transfer Entropy Approach for the Assessment of the Impact of Inspiratory Muscle Training on the Cardiorespiratory Coupling of Amateur Cyclists. *Front. Physiol.* **2020**, *11*, 134. [CrossRef]
12. de Abreu, R.M.; Cairo, B.; Porta, A. On the significance of estimating cardiorespiratory coupling strength in sports medicine. *Front. Netw. Physiol.* **2023**, *2*, 1114733. [CrossRef]
13. Kapidžić, A.; Platiša, M.M.; Bojić, T.; Kalauzi, A. RR interval-respiratory signal waveform modeling in human slow paced and spontaneous breathing. *Respir. Physiol. Neurobiol.* **2014**, *203*, 51–59. [CrossRef]
14. Bernardi, L.; Porta, C.; Gabutti, A.; Spicuzza, L.; Sleight, P. Modulatory effects of respiration. *Auton. Neurosci.* **2001**, *90*, 47–56. [CrossRef] [PubMed]
15. Spyer, K.M. Annual review prize lecture. Central nervous mechanisms contributing to cardiovascular control. *J. Physiol.* **1994**, *474*, 1–19. [CrossRef] [PubMed] [PubMed Central]
16. Feldman, J.L.; McCrimmon, D.R.; Morrison, S.F. Neural Control of Respiratory and Cardiovascular Functions. In *Fundamental Neuroscience*, 4th ed.; Elsevier Inc.: Amsterdam, The Netherlands, 2013; pp. 749–766. [CrossRef]
17. Eckberg, D.L. The human respiratory gate. *J. Physiol.* **2003**, *548 Pt 2*, 339–352. [CrossRef] [PubMed]
18. Grossman, P.; Taylor, E.W. Toward understanding respiratory sinus arrhythmia: Relations to cardiac vagal tone, evolution and biobehavioral functions. *Biol. Psychol.* **2007**, *74*, 2–285. [CrossRef] [PubMed]

19. Russo, M.A.; Santarelli, D.M.; O'Rourke, D. The physiological effects of slow breathing in the healthy human. *Breathe* **2017**, *13*, 298–309. [CrossRef] [PubMed] [PubMed Central]
20. Bernardi, L.; Valle, F.; Leuzzi, S.; Rinaldi, M.; Marchesi, E.; Falcone, C.; Martinelli, L.; Viganó, M.; Finardi, G.; Radaelli, A. Non-respiratory components of heart rate variability in heart transplant recipients: Evidence of autonomic reinnervation? *Clin. Sci.* **1994**, *86*, 537–545. [CrossRef] [PubMed]
21. Tafil-Klawe, M.; Trzebski, A.; Klawe, J. Contribution of the carotid chemoreceptor reflex to the mechanism of respiratory sinus arrhythmia in young healthy and hypertensive humans. *Acta Physiol. Pol.* **1985**, *36*, 59–64.
22. Yasuma, F.; Hayano, J. Respiratory sinus arrhythmia: Why does the heartbeat synchronize with respiratory rhythm? *Chest* **2004**, *125*, 683–690. [CrossRef] [PubMed]
23. López-Barneo, J.; Ortega-Sáenz, P.; González-Rodríguez, P.; Fernández-Agüera, M.C.; Macías, D.; Pardal, R.; Gao, L. Oxygen-sensing by arterial chemoreceptors: Mechanisms and medical translation. *Mol. Aspects. Med.* **2016**, *47–48*, 90–108. [CrossRef]
24. Karemaker, J.M. Counterpoint: Respiratory sinus arrhythmia is due to the baroreflex mechanism. *J. Appl. Physiol.* **2009**, *106*, 1742–1743; discussion 1744. [CrossRef] [PubMed]
25. Riedl, M.; Müller, A.; Kraemer, J.F.; Penzel, T.; Kurths, J.; Wessel, N. Cardio-respiratory coordination increases during sleep apnea. *PLoS ONE* **2014**, *9*, e93866. [CrossRef] [PubMed] [PubMed Central]
26. Krause, H.; Kraemer, J.F.; Penzel, T.; Kurths, J.; Wessel, N. On the difference of cardiorespiratory synchronisation and coordination. *Chaos* **2017**, *27*, 9. [CrossRef] [PubMed]
27. Hayano, J. Introduction to heart rate variability. In *Clinical Assessment of the Autonomic Nervous System*; Iwase, S., Hayano, J., Orimo, S., Eds.; Springer: Tokyo, Japan, 2017; pp. 109–127. [CrossRef]
28. Cover, T.M.; Thomas, J.A. Entropy, Relative Entropy, and Mutual Information. In *Elements of Information Theory*; Cover, T.M., Thomas, J.A., Eds.; Wiley: Hoboken, NJ, USA, 2005; pp. 13–55. [CrossRef]
29. Akaike, H. A new look at the statistical model identification. *IEEE Trans. Autom. Control* **1974**, *19*, 6–723. [CrossRef]
30. Hayano, J.; Yasuma, F. Hypothesis: Respiratory sinus arrhythmia is an intrinsic resting function of cardiopulmonary system. *Cardiovasc. Res.* **2003**, *58*, 1–9. [CrossRef] [PubMed]
31. Triedman, J.K.; Perrott, M.H.; Cohen, R.J.; Saul, J.P. Respiratory sinus arrhythmia: Time domain characterization using autoregressive moving average analysis. *Am. J. Physiol.* **1995**, *268 Pt 2*, H2232–H2238. [CrossRef]
32. Saul, J.P.; Berger, R.D.; Chen, M.H.; Cohen, R.J. Transfer function analysis of autonomic regulation. II. Respiratory sinus arrhythmia. *Am. J. Physiol.* **1989**, *256*, H153–H161. [CrossRef] [PubMed]
33. Lewis, G.F.; Furman, S.A.; McCool, M.F.; Porges, S.W. Statistical strategies to quantify respiratory sinus arrhythmia: Are commonly used metrics equivalent? *Biol. Psychol.* **2012**, *89*, 2–364. [CrossRef]
34. Spyer, K.M.; Lambert, J.H.; Thomas, T. Central nervous system control of cardiovascular function: Neural mechanisms and novel modulators. *Clin. Exp. Pharmacol. Physiol.* **1997**, *24*, 743–747. [CrossRef] [PubMed]
35. Tzeng, Y.C.; Larsen, P.D.; Galletly, D.C. Mechanism of cardioventilatory coupling: Insights from cardiac pacing, vagotomy, and sinoaortic denervation in the anesthetized rat. *Am. J. Physiol. Heart Circ. Physiol.* **2007**, *292*, H1967–H1977. [CrossRef]

MDPI AG
Grosspeteranlage 5
4052 Basel
Switzerland
Tel.: +41 61 683 77 34

Entropy Editorial Office
E-mail: entropy@mdpi.com
www.mdpi.com/journal/entropy

www.ingramcontent.com/pod-product-compliance
Lightning Source LLC
Chambersburg PA
CBHW041341210926
51850CB00013B/2947
9783725844494